Springer Theses

Recognizing Outstanding Ph.D. Research

For further volumes:
http://www.springer.com/series/8790

Aims and Scope

The series "Springer Theses" brings together a selection of the very best Ph.D. theses from around the world and across the physical sciences. Nominated and endorsed by two recognized specialists, each published volume has been selected for its scientific excellence and the high impact of its contents for the pertinent field of research. For greater accessibility to non-specialists, the published versions include an extended introduction, as well as a foreword by the student's supervisor explaining the special relevance of the work for the field. As a whole, the series will provide a valuable resource both for newcomers to the research fields described, and for other scientists seeking detailed background information on special questions. Finally, it provides an accredited documentation of the valuable contributions made by today's younger generation of scientists.

Theses are accepted into the series by invited nomination only and must fulfill all of the following criteria

- They must be written in good English
- The topic of should fall within the confines of Chemistry, Physics and related interdisciplinary fields such as Materials, Nanoscience, Chemical Engineering, Complex Systems and Biophysics.
- The work reported in the thesis must represent a significant scientific advance.
- If the thesis includes previously published material, permission to reproduce this must be gained from the respective copyright holder.
- They must have been examined and passed during the 12 months prior to nomination.
- Each thesis should include a foreword by the supervisor outlining the significance of its content.
- The theses should have a clearly defined structure including an introduction accessible to scientists not expert in that particular field.

Chen Davidovich

Targeting Functional Centers of the Ribosome

Doctoral Thesis accepted by
Weizmann Institute of Science (WIS), Rehovot, Israel

 Springer

Author
Dr. Chen Davidovich
Department of Structural Biology
Weizmann Insitute of Science (WIS)
Herzl Street
76100 Rehovot
Israel
e-mail: chend@weizmann.ac.il

Supervisor
Prof. Ada Yonath
Department of Structural Biology
Weizmann Insitute of Science (WIS)
Herzl Street
76100 Rehovot
Israel
e-mail: ada.yonath@weizmann.ac.il

ISSN 2190-5053

e-ISSN 2190-5061

ISBN 978-3-642-16930-4

e-ISBN 978-3-642-16931-1

DOI 10.1007/978-3-642-16931-1

Springer Heidelberg Dordrecht London New York

© Springer-Verlag Berlin Heidelberg 2011

Cover design: eStudio Calamar, Berlin/Figueres

Printed on acid-free paper

Springer is part of Springer Science+Business Media (www.springer.com)

Parts of this thesis have been published in the following journal articles:

C. Davidovich, M. Belousoff, I. Wekselman, T. Shapira, M. Krupkin, E. Zimmerman, A. Bashan, A. Yonath. "The Proto-Ribosome: An Ancient Nano-machine for Peptide Bond Formation" *Isr. J. Chem.*, (2010) 50 (1), pp. 29–35.

M. J. Belousoff, C. Davidovich, E. Zimmerman, Y. Caspi, I. Wekselman, L. Rozenszajn, T. Shapira, O. Sade-Falk, L. Taha, A. Bashan, M. S. Weiss, A. Yonath. "Ancient machinery embedded in the contemporary ribosome." *Biochem. Soc. Trans.*, (2010) 38 (2), pp. 422–7.

T. Auerbach, I. Mermershtain, C. Davidovich, A. Bashan, M. Belousoff, I. Wekselman, E. Zimmerman, L. Xiong, D. Klepacki, K. Arakawa, H. Kinashi, A. Mankin and A. Yonath, "The structure of ribosome-lankacidin complex reveals ribosomal sites for synergistic antibiotics", *Proc Natl Acad Sci U S A*, (2010) Epub since Jan 11.

C. Davidovich, M. Belousoff, A. Bashan and A. Yonath, "The evolving ribosome: from non-coded peptide bond formation to sophisticated translation machinery", *Res. Microbiol.*, (2009) 160 (7), pp. 487–92.

T. Auerbach, I. Mermershtain, A. Bashan, C. Davidovich, H. Rosenberg, D. H. Sherman, and A. Yonath, "Structural basis for the antibacterial activity of the 12-membered-ring mono-sugar macrolide methymycin", *Biotechnologia*, (2009) 84, pp. 24–35.

C. Davidovich, A. Bashan, A. Yonath, "Structural basis for cross-resistance to ribosomal PTC antibiotics", *Proc Natl Acad Sci U S A*, (2008) 105 (52), pp. 20665–70.

I. Wekselman, C. Davidovich,[1] I. Agmon, E. Zimmerman, H. Rosenberg, A. Bashan, R. Berisio and A. Yonath, "Ribosome's mode of function: myths, facts and recent results", *J Pept Sci*, (2008) 15 (3), pp. 122–30.

C. Davidovich, A. Bashan, T. Auerbach-Nevo, R. D. Yaggie, R. R. Gontarek, A. Yonath, "Induced-fit tightens pleuromutilins binding to ribosomes and remote interactions enable their selectivity", *Proc Natl Acad Sci U S A*, (2007) 104 (11), pp. 4291–96.

---

[1] Wekselman and Davidovich contributed equally.

Supervisor's Foreword

Ribosomes, the key players in the translation process, are universal ribozymes performing two main tasks: decoding genetic information and polymerizing amino acids. Hundreds of thousands of ribosomes operate in each living cell due to the constant degradation of proteins through programmed cell death, which is matched by simultaneous production of proteins. For example, quickly replicating cells, e.g. liver cells, may contain a few million ribosomes. Even bacterial cells may contain to 100,000 ribosomes during their log period. Other constituents are the mRNA chains, produced by the transcription of the segments of the DNA that should be translated, which carry the genetic information to the ribosomes, and tRNA molecules bring the cognate amino acids to the ribosome. To increase efficiency, a large number of ribosomes act simultaneously as polymerases, synthesizing proteins by one-at-a-time addition of amino acids to a growing peptide chain, while translocating along the mRNA template, and producing proteins on a continuous basis at an incredible speed, namely up to 20 peptide bonds per second.

Ribosomes are giant assemblies composed of many different proteins (r-proteins) and long ribosomal RNA (rRNA) chains. Among these, the RNA moieties perform the two ribosomal main functions. The ratio of rRNA to r-proteins ($\sim 2{:}1$) is maintained throughout evolution, except in mitochondrial ribosome (mitoribosome) in which $\sim$ half of the bacterial rRNA is replaced by r-proteins. Nevertheless, the active regions are almost fully conserved in all species. In all organisms ribosomes are built of two subunits, which associate to form the functionally active ribosomes. In prokaryotes, the small subunit, denoted as 30S, contains an RNA chain (16S) of ~ 1500 nucleotides and ~ 20 different proteins. The large subunit (50S in prokaryotes) has two RNA chains (23S and 5S RNA) of about 3000 nucleotides in total, and different >31 proteins. The available three dimensional structures of the bacterial ribosome and their subunits show that in each of the two subunits the ribosomal proteins are entangled within the complex rRNA conformation, thus maintaining a striking dynamic architecture that is ingeniously designed for their functions: precise decoding; substrate mediated peptide-bond formation and efficient polymerase activity.

While the elongation of the nascent chain proceeds, the two subunits perform cooperatively while the tRNA molecules are the non-ribosomal entities combine the two subunits. The small subunit provides the path along which the mRNA progresses, the decoding center and the mechanism controlling translation fidelity. Translation initiation is the rate-limiting step of the entire process. It starts by the correct selection and placement of the mRNA reading frame, and proceeds through a tightly regulated decoding at the P-site. The large subunit contains the site for the main ribosomal catalytic function, namely polymerization of the amino acids and provides the dynamic protein exit tunnel. Simultaneously with the advancement of the mRNA along the path in the small subunit, peptide bonds are being formed. This inherently dynamic process requires small and large-scale motions of the ribosomal substrates coupled to conformational rearrangements of its components and substrates. The nascent proteins progress along a dynamic tunnel and emerge from the large subunit into a shelter formed by ribosome-bound trigger-factor, acting as a chaperone preventing aggregation and misfolding. The current consensus view is consistent with ribosomal positional catalysis, namely providing suitable stereochemistry for peptide bond formation accompanied by appropriate geometrical means for substrate mediated catalysis, and not by acid/base mechanism.

Chen Davidovich joined our group after the molecular structures of the two bacterial ribosomal subunits had been determined and their functional regions identified and localized. In his PhD thesis Chen aimed to reveal the structural basis for the catalytic function of the ribosome as a polymerase, namely as the cellular machine that forms peptide bonds successively, thus elongating the nascent chains. When he started, this approach was rather neglected since most ribosomologists assumed that understanding the formation of a single peptide bond was sufficient to describe the ribosomal function, and hence disregarded the processivity and elongation.

Chen investigated in detail the mode of function, inhibition, and evolution of the Peptidyl Transferase Center (PTC) by a combination of X-ray crystallography, biochemistry, molecular biology and theoretical studies. Among his major achievements was the determination of the crystal structures of complexes of ribosomal particles with antibiotics that target the PTC. These are clinically useful, despite the extremely high conservation of this region in all forms of life. Chen revealed the modes of action of these antibiotics, thus shedding light on the discrimination principles between ribosomes of eubacterial pathogens and the eukaryotic hosts that are based on remote interaction networks and induced fit. In parallel, he investigated the mechanisms acquiring resistance to antibiotics. Furthermore, by thorough comparative structural and genetic analyses, he determined the structural basis for crossresistance between all of the clinically useful PTC-binding antibiotics.

Chen also attempted to elucidate the origin of the ribosome. He was focused on the ribosomal substructure that may represent the minimal entity capable of performing peptide bond formation, namely the proto-ribosomes. Remarkably, despite the ribosome asymmetric structure, in all of the structures determined so far the PTC and its environs are situated within a highly conserved region of internal structural symmetry that connects all ribosomal functional centers involved in amino-acid polymerization, hence can serve as the central signaling feature. The high level of conservation of the symmetrical region suggests that the modern ribosome evolved from a simpler entity that can be described as a proto-ribosome that was formed by gene fusion or gene duplication and contained a pocket confined by two self folded RNA chains, which associated to form a pocket like dimer with functional capabilities. As RNAs can act as gene-like molecules coding for their own reproduction, it is conceivable that the surviving pockets became the templates for the ancient ribosomes. In a later stage these RNA genes underwent initial optimization to produce a more defined rather stable pocket, in which each of the two halves was further optimized for its task in distinction between the amino acid and the growing peptidyl sites, so that their sequences evolved differently. Indeed, the preservation of the three-dimensional structure of the two halves of the ribosomal frame regardless of their sequence demonstrates the superiority of rigorous functional requirements over sequence preservations.

Two symmetry-related RNA entities seem to have not only the capability to self fold, but can also undergo self dimerization. Based on the assumption that dimerization is the minimal requirement for formation of catalytic active pockets, Chen designed and synthesized RNA chains corresponding to the sequence of this pocket in the contemporary ribosome and determined their tendency to dimerize. He found, surprisingly, a marked preference of specific sequences to dimerize, whereas other, very similar RNA chains did not. This sequence preference for self dimerization, indicate that the principles of 'survival of the fittest' may have played a major role on the molecular level, although these terms are usually used for species, hence suggesting pre Darwinian Darwinisim.

Chen's thesis has already led to attempts at improving the PTC antibiotics, performed locally as well as by companies. Additionally, a significant effort is being made by two graduate students and a postdoctoral fellow, to try to design a functional construct mimicking a functional proto-ribosome.

December 2010   Ada Yonath

Contents

Chapter 1
Introduction

The ribosome is a universal cellular organelle that translates the genetic code into proteins. It is an assembly of ribosomal proteins (r-proteins) and ribosomal RNA (rRNA) chains that facilitates the reaction of amino acid polymerization into proteins, according to specific genetic information that is encoded in the cell genome and transferred to the ribosome by messenger RNA (mRNA). The ribosome is responsible for the identification of the correct genetic reading frame and translation initiation. It ensures robust fidelity when translating the genetic code by selecting the proper transfer RNA (tRNA) molecules that carry the cognate amino acids to be incorporated into the elongating polypeptide chain. It provides the framework for peptide bonds formation and smooth elongation of the newly synthesized proteins, including co-translational folding, chaperoning by ribosome-associated factors and the exit tunnel. When stop cordon arrive at the decoding site, the ribosome utilizes recycling and release factors to execute translation termination (reviewed in [1–5]).

The ribosome is composed of two counterparts; a large subunit, with a typical sedimentation coefficient of 50S in prokaryotes, and a small subunit, with a typical sedimentation coefficient of 30S in prokaryotes. The two subunits associate to initiate the process of protein biosynthesis and dissociate as the process terminates. Throughout protein biosynthesis the associated ribosomal subunits facilitate different tasks: the small subunit is responsible for initiation, identification of the correct reading frame and encoding of the genetic code. The main chemical reaction of protein biosynthesis, peptide bond formation, occurs in the large subunit, which ensures smooth polymerization, channeling of the nascent chain through the exit tunnel and facilitate folding by the aid of associating factors.

The first phase in protein biosynthesis is the formation of the initiation complex: small ribosomal subunit, mRNA, initiator tRNA (in bacteria fMet-tRNA^{fMet}) that binds to the AUG start codon and the initiation factors. Next, association of the two subunits results with an active ribosomal complex. The amino acids are carried into the ribosome by tRNA molecules, aminoacylated to its 3′ end (aminoacyl tRNA), which consist of a universally conserved CCA sequence. The

C. Davidovich, *Targeting Functional Centers of the Ribosome*, Springer Theses, DOI: 10.1007/978-3-642-16931-1_1, © Springer-Verlag Berlin Heidelberg 2011

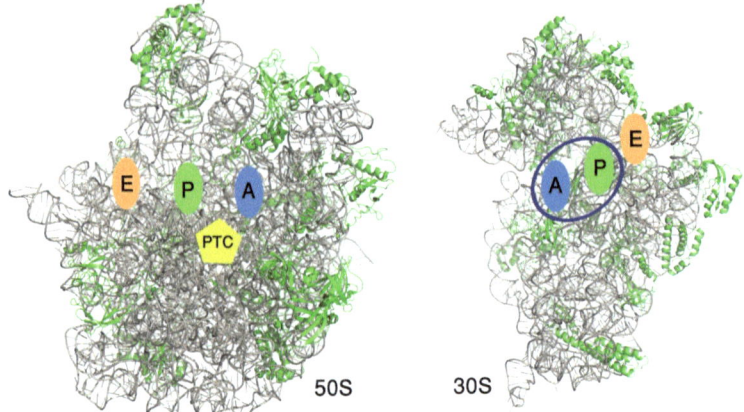

Fig. 1.1 Front views (the subunit interfaces) of the two eubacterial ribosomal subunits. Shown are the large ribosomal subunit from *Deinococcus radiodurans* (50S, *left panel*) and the small ribosomal subunit from *Thermus thermophilus* (30S, *right panel*). Ribosomal RNA is shown in gray, proteins colored green. Blue, green and orange ellipsoids marked A, P, and E, respectively, designate the three tRNA binding sites within the small subunit (namely, the approximate positions of the anticodon loops) and the large subunit (namely, the approximate positions of the tRNA acceptor stems). The yellow pentagon indicates the approximate location of the PTC in the large subunit and blue ellipsoid surrounding the A and P sites within the small ribosomal subunit indicating for the decoding center

ribosome contains three tRNA binding sites: A (aminoacyl tRNA), P (peptidyl tRNA) and E (exit) (see Fig. 1.1). The first aminoacyl tRNA 3' end binds to the large subunit at the P-site. This is followed by the accommodation of the A-site tRNA so that both anticodon loops maintain the Watson–Crick and Wobble base pairs for codon–anticodon interactions with the mRNA codon that is bound at the P- and the A-site at the decoding center, within the 30S subunit. While the proper stereochemistry is achieved between both substrates peptide bond is being formed within the PTC void, following translocation of the tRNA from the A- to the P-site and the ribosome is ready to receive new amino acylated tRNA substrate. The deacylated P-site tRNA acts as a leaving group and moves towards the E-site, on its way out from the ribosome. The newly synthesized protein is growing into the exit tunnel at the 50S subunit, adjacent to the PTC.

Efforts for ribosomes crystallization began three decades ago [6], but these enormously complicated crystallographic studies became fruitful only at the end of the last decade, yielding the three dimensional structures of ribosomal particles from three microorganisms. These include the high resolution structures of the small ribosomal subunit (30S) of the thermophilic bacteria *T. thermophilus* (T30S) [7, 8], the large ribosomal subunit (50S) of the halophilic archaeon *H. marismortui* (H50S) [9], the mesophilic eubacterium *D. radiodurans* (D50S) [10]. Half a decade later the structures of the empty 70S ribosome from *E. coli* (E70S) [11] and of the 70S ribosomal complexes from *T. thermophilus* (T70S) with mRNA and tRNA molecules [12, 13].

Based on the structure of H50S it has been proposed that three specific rRNA nucleotides within the PTC provide chemical catalysis for peptide bond formation [14]. Later it was shown that the PTC and its upper rim provides the means for accurate positioning of the tRNA, in an orientation required for the nucleophilic attack on the P-site carbonyl carbon [15, 16]. Based on this finding, together with results of genetic, biochemical and theoretical studies (reviewed in [4, 17–21]) it has been suggested that the ribosome provides positional catalysis to facilitate peptide bond formation.

A reliable structural approach to gain knowledge regarding ribosomal function and structure is by using X-ray crystallography techniques on ribosomal particles complexed with different antibiotics. Ribosomal antibiotics typically bind to functional centers and hamper ribosomal activity. These relatively small compounds enable feasible soaking protocols or co-crystallization procedures that were found to yield well diffracting crystals of the ribosome/antibiotic complex. Such studies shed light on the role of the small ribosomal subunit [2, 22, 23], revealed some of the fundamental characteristics of the PTC [24–29] and the polypeptide exit tunnel [29–35].

Furthermore, careful analysis of complexes of the large subunit with substrate analogs revealed a universal sizable symmetrical region of 180 nucleotides within the otherwise asymmetric ribosome. This symmetrical region is located in and around the PTC, at the heart of the ribosome. This region provides the structural elements allowing for peptide bond formation and tRNA translocation. It connects, directly or via its extensions, all the ribosomal functional sites involved in amino acid polymerization [16]. Its key role in protein biosynthesis led to the suggestion that the ribosome evolved by gene fusion or gene duplication [15].

1.1 Ribosomal Antibiotics: Function, Activity and Selectivity

Having a fundamental role in cell viability and although being highly conserve, the ribosome has some diverge genetic characteristics that distinguish between bacteria and eukaryotes. Thus, the ribosome is a major target for clinically useful antibiotics. By targeting functional centers of the ribosome, these drugs compete with substrates and factors binding.

Crystal structures of ribosome–antibiotic complexes revealed different conformational states of functional sites, thus are described as a "snapshot" of a "frozen" conformation. For instance, the antibiotic Synercid® composed of two synergistically active streptogramin compounds dalfopristin (S_A) and quinupristin (S_B), each of which is a poor antibiotic agent. They bind to the PTC and the macrolide binding site close to the tunnel entrance, respectively [27]. Quinupristin partially blocks the exit tunnel and dalfopristin stabilizes a swung conformation of U2585 (nucleotides are numbered according to *E. coli* system throughout, unless otherwise mentioned), which is flipped by 180° with respect to its location in the D50S apo structure. Such enormous conformational rearrangement reflects the high flexibility of this nucleotide that was suggested to serve as an anchor for the tRNA A- to P-site rotatory motion [36]. This simple example emphasizes the possible usage of ribosome inhibitors as a valuable tool for functional studies.

Ribosomal antibiotics must not only bind with high affinity, but their clinical usage requires high selectivity, namely; having the ability to target and inhibit the bacterial ribosome, without hampering the ribosome of the eukaryotic host. This is usually achieved as a result of minor differences between ribosomal functional sites in bacteria and eukaryotes. An example is the macrolide class of antibiotics, which all share interactions with nucleotide A2058 [29]. This nucleotide is a conserved adenine in eubacteria and a conserved guanine in archaea and eukaryotes [37], hence provides selectivity for these compounds. Indeed, the mutation A2058G in bacteria leads to macrolides resistance [38] while the mutation G2058A in the archaeon *H. marismortui* permits macrolides binding to the ribosome of this microorganism [34], which naturally is not sensitive to these antibiotics.

1.2 Pleuromutilins: A Class of PTC Ribosomal Antibiotics for Clinical Use

As a result of the dramatic increase in antibiotic resistance among pathogenic strains, which causes a major clinical threat in recent years, the arsenal of efficient antibacterial drugs decreases. A promising strategy for reducing the magnitude of this problem is the introduction of antibiotics from novel classes that so far have not been used clinically and have a unique mode of action. An example is the pleuromutilin family.

Pleuromutilin was discovered in 1951 as a natural product of the fungi *Pleurotus mutilus* (nowadays *Clitopilus scyphoides*) [39] (see Fig. 1.2 for chemical formula). Although the natural drug exhibited reasonable, however not excellent in vitro activity against *Staphylococcus aureus*, there was no advantage in using it rather than any other available antibiotic; hence it was abandoned for the next two decades. The discovery made in the seventies, that semi-synthetic pleuromutilin derivatives may show elevated activity over a broad spectrum [40] paved the way for further development of this new class of antibacterial agents, initially used as veterinary drugs, namely tiamulin and valnemulin (see Fig. 1.2 for chemical formula). During the early 1980s, extensive efforts were made in order to formulate azamulin (TDM-85,530) (see Fig. 1.2 for chemical formula), a pleuromutilin derivative for clinical use in human [41–43]. Although this compound is active against a large variety of clinical isolates, including erythromycin- and tetracycline-resistant strains, it strongly inhibits cytochrome P-450 [44], thus was not suitable for clinical use and never progressed beyond phase I clinical trials [45]. However, continuous efforts to further develop pleuromutilin antibiotics yielded several novel semisynthetic compounds, some of which with increased activity and reduced

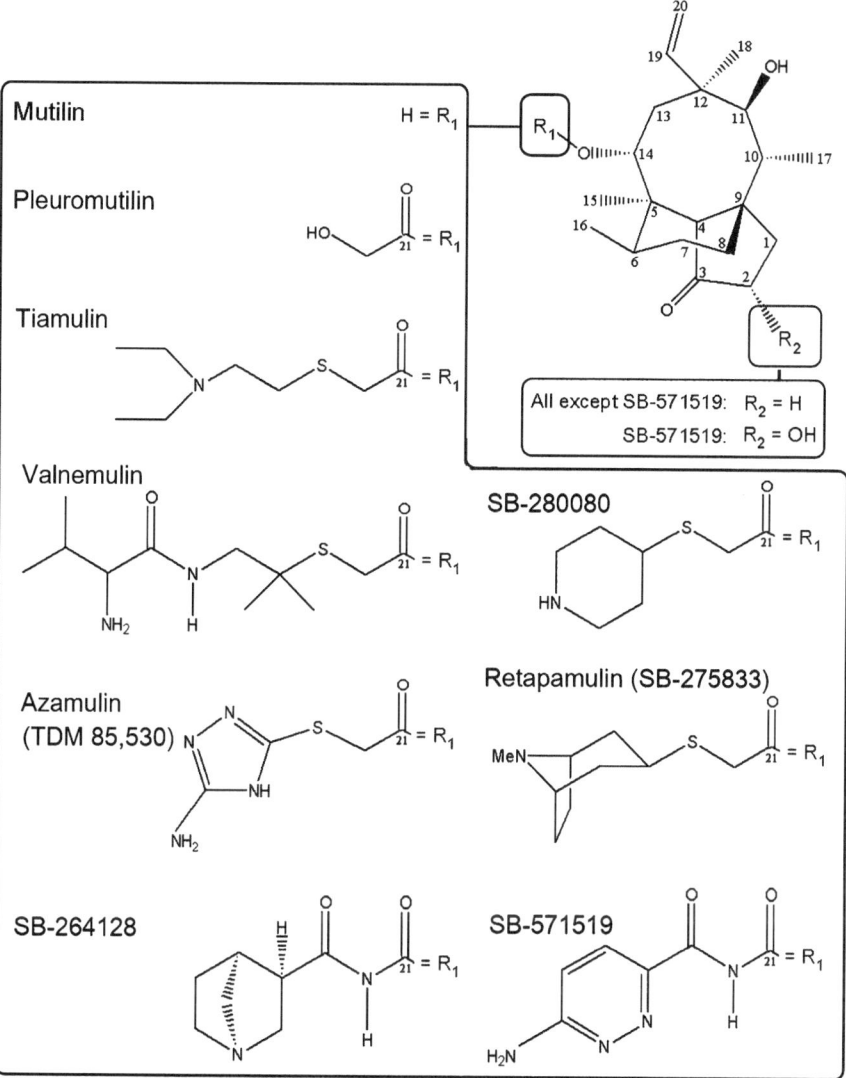

Fig. 1.2 Chemical formula of selected pleuromutilin derivatives. Typically, the tricyclic mutilin core is conserved among pleuromutilin derivatives. Notable is the variability of the C14 extension (R1 in the large box). *Large box*: previously studied pleuromutilins (*left*), pleuromutilins characterized within the present study (*right*), mutilin and the natural pleuromutilin (*upper left*), acyl-carbamate semisynthetic derivatives (*bottom*). Notable is the conserved C21 carbonyl that is crucial for pleuromutilins activity (see discussion). *Upper right, small box*: R2, the acyl-carbamate derivative SB-571519 includes an additional hydroxyl that substituted on C2

metabolism. Among these, a few derivatives are currently in the industries' pipelines, and the most advanced—retapamulin (SB-275833) [46–52], was already issued for clinical use in human [53–56].

The first potent group of semisynthetic pleuromutilin is of the C(14)-sulfanyl-acetate derivatives. This group includes the recently formulated drug retapamulin, the commonly used veterinary drugs tiamulin and valnemulin, and other potent compounds, such as azamulin and SB-280080 (see Fig. 1.2 for chemical formula). This group of compounds is characterized by high anti-bacterial activity against various clinically relevant Gram negative bacterial strains (for review see [57]). However, C(14)-sulfanyl-acetate derivatives undergo rapid metabolism by cytochrome P-450 [57, 58], thus characterized by limited oral bioavailability.

An additional group of pleuromutilins with good antibacterial activity are the C(14)-acyl-carbamate derivatives. This group, which includes SB-571519 and SB-264128 (Fig. 1.2), are slightly less potent than the sulfanyl-acetate derivatives, but it is characterized as having a reduced metabolism with elimination rate of down to tenth in respect to the C(14)-sulfanyl-acetate derivatives [58]. As this is a major advantage from a clinical point of view, these compounds were originally aimed for oral treatment.

Pleuromutilins block peptide bond formation by interfering with substrate binding both at the A- and the P-sites of the ribosomal PTC [59, 60]. It was found that the drug competes with 3′-terminues tRNA analogues, like puromycin and CpCpA oligonucleotide, and with the antibiotic chloramphenicol [61] that binds at the PTC A-site [29]. However, once elongation process has begun; translation will not be effected by the drug [62]. Consistently, the detailed crystal structure of the large ribosomal subunit from *D. radiodurans* in complex with the pleuromutilin class antibiotic tiamulin indicates binding to the PTC [60], overlapping the binding sites of the A- and P-substrates.

Yet, until the point where my study was launched, no extensive comparative structural analysis of pleuromutilin antibiotics has been conducted, and the only structural information describing this class of antibiotics was heavily relayed on, and biased by, a single antibiotic designated for veterinary use, namely tiamulin. Under these circumstances; general mechanisms for drug binding and antibiotic resistance describing this class were vague and details providing hints for further drug design were lacking. Moreover, to that point no mechanism for drug selectivity was known. Within my research, as describes herein, I carried out the first comparative structural study of this class of antibiotics. Herein I describe general and detailed structural mechanisms for binding, resistance and selectivity of pleuromutilin antibiotics.

1.3 Structural Basis for Cross-Resistance Between Ribosomal PTC Antibiotics

Among the various classes of clinically useful antibiotics that target the ribosome, five classes target the PTC, namely phenicols, lincosamines, streptogramins A, pleuromutilins and oxazolidinones. These drugs bind to the PTC of bacterial ribosomes with high affinity and great specificity, thus hamper ribosomal activity, resulting in the loss of bacterial cell vitality, without significant effect on the eukaryotic host.

Cross-resistance appears when two or more drugs share binding sites with overlapping positions. According to crystal structures of the large ribosomal subunits from the eubacteria *D. radiodurans* (D50S) and the archaeon *H. marismortui* (H50S) that were complexed with various PTC binding antibiotics [26–29, 60, 63] this is the case for all the five classes of antibiotics mentioned above.

All of these antibiotics were shown to bind with overlapping sites in the PTC A-site, in the vicinity of nucleotide U2504. The only exception was the case of chloramphenicol that was shown in two different locations in D50S and H50S. While in D50S the drug was shown in the A-site, with overlapping positions with all other PTC antibiotics, the H50S/chloramphenicol complex indicating a controversial binding site at the tunnel entrance, possibly because of the extremely high antibiotic concentration of 20 mM that was used in that experiment and the "lower affinity of the archaeal ribosome A-site crevice for chloramphenicol consistent with archaeal ribosomes being resistant to inhibition by chloramphenicol" [26].

In recent years few indications accumulated indicating cross-resistance between some PTC antibiotics [64–66]. In addition, nucleotide U2504 was suggested to be an impotent intermediate that carries conformational changes induced by mutations in remote nucleotides and amino acids that are involved in tiamulin resistance [67]. By performing a comprehensive structural study on pleuromutilin antibiotics [63] and comparing with crystal structure of tiamulin/D50S complex [60], we have suggested that similar mutations are likely to affect other members of this class. This finding was based on the discovery that all the studied pleuromutilins derivatives bind with their common tricyclic core at overlapping positions in a tight binding pocket that includes nucleotide U2504. Similar mechanism for resistance and selectivity was later suggested based on the analysis of the locations of resistance mutations to the archaeal and eukaryotic antibiotic anisomycin [25].

While PTC antibiotics from five different classes can be used clinically and although in the recent years valuable information regarding resistance mechanism for each class of these drugs was obtained by structural and biochemical studies, not much has been done in order to derive general properties of these classes. Such analysis can predict common patterns for resistance mechanisms. As this issue has great value in light of the emergence evidences for cross-resistance between PTC antibiotics of different classes [64–66, 68, 69], I conducted a cooperative study. Herein I present a comprehensive structural analysis aimed at reveling common characteristics that govern the mechanism of drug resistance in the PTC. This was done by analyzing the spatial distribution of nucleotides that affect resistance to PTC antibiotics, as well by suggesting a structure based mechanism by which remote mutations alter the binding site of these drugs.

1.4 Oligonucleotides as Ribosomal Inhibitors and as Tools for Structural and Functional Study

Long before the emergence of the high resolution crystal structures of the ribosome short oligonucleotides were used to structurally and functionally characterize it. Short DNA oligonucleotides, down to six bases long, were used to probe rRNA accessibility. Nucleotides of this length are sufficient to hybridize into DNA–RNA double strands, hence can target accessible functional sites in the 16S rRNA during initiation complex formation [70]. Such nucleotides were also exploited to probe 16S region involved in ribosomal subunits association [71], to map the decoding center in the presence and the absence of polyU and tRNA^{Phe} [70], and to probe the interactions of tRNA with the 23S rRNA [72, 73]. In another study a DNA probe was used to locate a known RNA sequence in the low resolution EM map of the ribosome, using 12 bases long oligonucleotide with a terminal modification designed for immune electron microscopy [74]. This probe was designed to hybridize to the 16S rRNA sequence 520–531, near the decoding site (Fig. 1.1). The correct identification of the decoding center within these EM maps was confirmed when the high resolution structures of the small ribosomal subunit appeared ten years later [7, 8]. Another technique exploited DNA antisense oligonucleotides that were chemically modified with a photo-sensitive affinity label. Such probe enabled site specific cross-linking, thus identified specific neighborhoods. This method showed cross-linking of an antisense DNA complementary to bases 2497–2505 of domain V of the 23S rRNA to r-protein L3 [75], an observation that was later confirmed by the X-ray structure of the eubacterial large ribosomal subunit [10].

While the majority of these studies concentrated on rRNA sequences that are currently known to reside on accessible regions of the ribosomal subunits, a similar study concentrated on the PTC cavity [76]. In this study, a ^{32}P labeled DNA probes, 6–15 bases long, were used to probe rRNA accessibility in the presence of the ribosomal antibiotic chloramphenicol. This study demonstrated DNA oligonucleotides binding to ribosomal functional centers located in deep ribosomal cavities, as is currently known for the PTC.

The idea that short antisense oligonucleotide sequences can selectively inhibit the bacterial ribosome has already been confirmed. One of the first attempts for a rational design of oligonucleotide ribosomal inhibitor was based on the Shine–Dalgarno (SD) sequence. The SD sequence is a short sequence (AGGAGG) within the mRNA that signals for initiation of protein biosynthesis in bacteria [77]. It is located upstream of the first coding AUG, and its complimentary sequence (ACCUCCU) is located at the 3′ end of the 16S rRNA in the small subunit of the bacterial ribosome. The idea that specific recognition between a short mRNA sequence and the 16S rRNA is directly responsible for mRNA binding to the ribosome led to the suggestion that such short oligonucleotide sequence may selectively inhibit the bacterial ribosome [78]. In that work, Jayaraman and coworkers used short (3–7 bases) nonionic oligonucleotides (deoxy ribo oligonucleoside methylphosphonates) that were complimentary to the anti-SD sequence at the 16S rRNA. The most potent sequences were of 6 and 7 bases long that mimicked the SD sequence (AGGAGG and AGGAGGT, respectively). These two oligonucleotides sequences inhibited protein synthesis of in vitro transcription translation system with IC_{50}, namely the inhibitor concentration required to inhibit the system down to 50% of its observed inhibitor-free activity, of between 25 and 12.5 μM. Such IC_{50} values are lower than those previously measured for the standard ribosomal antibiotic tetracycline (30 μM), and just slightly higher than the IC_{50} values of the ribosomal antibiotics chloramphenicol (5 μM) and linezolide (2.5 μM) [79]. However, these oligonucleotides failed to inhibit the growth of intact *E. coli*, due to low permeability. A subject for similar studies was the α-sarcin domain of the 23S rRNA (2646–2674). This stem-loop contains 12 universally conserved nucleotides, and is among the most conserved sequences in the 23S rRNA. A hydrolysis of the rRNA backbone after nucleotide G2661 by the highly specific RNase α-sarcin blocks all elongation factor dependent reactions [80]. An antisense DNA oligonucleotide sequence complimentary to this region was found to inhibit ribosome activity both in eukaryotes [81] and prokaryotes [82]. However, effective inhibition of poly(Phe) synthesis was achieved only at low temperature of 25°C.

In another study a set of 12 overlapping antisense 2′-O-methyl 10-mer oligoribonucleotides were designed to target nucleotides 1485–1516 in the A-site of the small ribosomal subunit [83]. The binding and inhibition properties of these oligonucleotides were tested, and some of them were found to bind ribosomes with K_d of down to 30 nM, as well as with the ability to inhibit in vitro transcription–translation system with IC_{50} of as low as 10 μM. The approach of using predefined length antisense oligonucleotides as antisense probes is suitable for targeting relatively short and defined rRNA regions, but it can become more extensive if considering longer rRNA target sequences, and oligomers with various length.

In the studies described above, the antisense oligonucleotides were designed either according to the target rRNA sequence, or based on considerations regarding the two dimensional rRNA structure. In some cases, as described above, these attempts were successful. However, not all of them were of the same efficiency, and some were totally inefficient.

During the past two decades several theoretical methods were developed in order to predict RNA secondary structure, and its hybridization with antisense oligonucleotides (e.g. [84–88]). Such approaches may be suitable for single unbound RNA fragments, as mRNA, or in cases of well thermodynamically defined RNA structures, however are inappropriate for RNA–protein assembly as the ribosome, which includes a large inaccessible core, tens of incorporated proteins, and convoluted thermodynamics that is far from been understood.

Neither of the early antisense oligonucleotides that have been proved to serve as effective ribosome inhibitors [78, 81–83] and ribosome targeting probes [72, 73, 76] designed based on three dimensional structural information. On the other hand, coordinates of high resolution crystal structures of ribosomal complexes are available. Yet, when I started no three dimensional structure based strategy has been developed to derive effective antisense sequences that will bind to exposed

RNA elements of molecules or complexes with known structure, as the ribosome. Hence, I initiated a process for the development of a novel approach for structure-based design of antisense oligonucleotide sequences capable of binding to ribosomal exposed functional centers. Such oligonucleotides can potentially be used as inhibitors and also serve as the basis for future antibacterial drugs.

1.5 Minimal Ribosomal Components with PTC Structure and Function

Since the discovery made in the 1960s that the ribosome can catalyze the formation of peptide bonds between minimal substrates, e.g. puromycin and fMet-tRNA [89] or CAACAA-formyl-methionine [90], in a biochemical reaction called the "fragment reaction", efforts were made in order to search for the smallest set of ribosomal components that are capable of carrying out such reaction. During the early 1980s Schulze and Nierhaus found that the large ribosomal subunit per se possesses peptidyl transferase activity even in the presence of only the 23S rRNA and 15 r-proteins, out of its "34 distinct components" [91].

A decade later Noller showed that the large ribosomal subunit has catalytic activity after vigorous treatment with proteinase K and SDS [92]. Partial activity was also observed in reconstituted large subunits from T7 RNA polymerase transcribed 23S and 5S rRNA and r-proteins that were extracted from 50S subunit [93]. The smallest active sub-ribosomal particle that obtained in those studies included both 5S and 23S rRNA and 8 ribosomal proteins [94]. Yet, while the combination of several intact ribosomal component were shown to maintain some level of ribosomal activity, trials to obtain peptidyl transferase activity by designing minimal construct that will include only small portion of 23S rRNA from the vicinity of the PTC were unsuccessful [95–99].

The PTC, i.e. the core catalytic center and the site of peptide bond formation, is composed solely of highly conserved ribosomal RNA (rRNA). It is located at the heart of the contemporary ribosome in the midst of a symmetrical 'pocket-like' structure [15, 16, 100, 101] (Fig. 1.3). This is an unusual feature within the otherwise asymmetric ribosome. This region exists in all known high resolution crystal structures of ribosomal complexes [9–13] and its inner part displays resistance to mutations [102]. Its 180 nucleotides are divided into two distinct regions that are related by a pseudo two fold structural symmetry, irrespective of the nucleotide sequences.

This symmetrical 'pocket-like' region hosts the two ribosome's substrates, namely the amino acylated and the peptidyl tRNA molecules. The elaborate architecture of this structural element provides the framework for the ribosome catalytic activity, as it positions the ribosome's substrates in favorable stereochemistry for peptide bond formation [15, 16, 19, 100, 101, 103–105] and for substrate-mediated catalysis [17, 106–108]. Additionally, by encircling the PTC it

Fig. 1.3 The symmetrical region within the large ribosomal subunit. *Top*: The A-region is shown in blue, the P-region in green, and the substrate position is indicated by the orange object. *Bottom*: the GNRA tetraloop and the A-minor interaction between the A-region (blue) and the P-region (green)

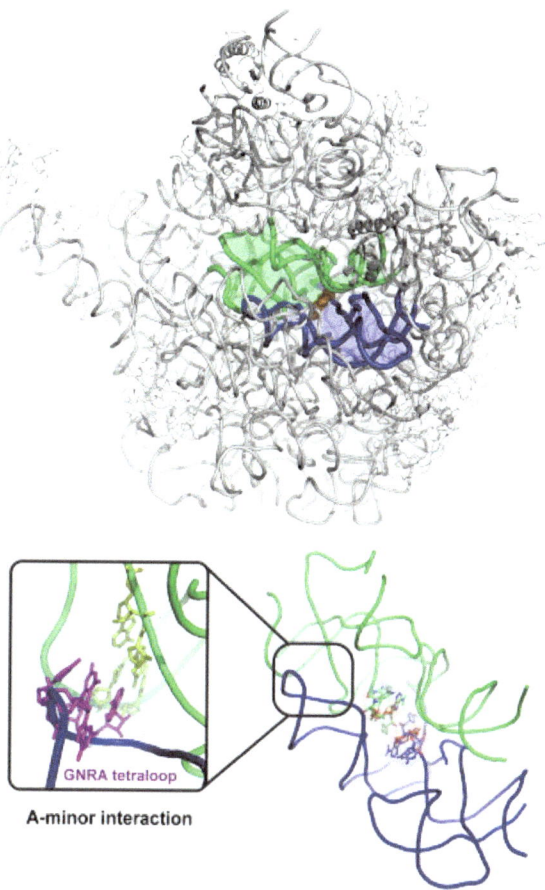

confines the void required for the motions involved in the translocation of the tRNA 3′ end, which, in turn, is necessary for successive peptide bond formation, enabling the amino acid polymerase activity of the ribosome [15, 17, 100, 101]. The rRNA entities of the symmetrical region possess the common stem-elbow-stem (SES) structural motif, and are hypothesized to be capable of self-assembly and dimerization [100, 101].

Based on the above observations we have proposed that the ancient machinery that could form peptide bonds was made exclusively from RNA molecules that self-assembled to create the ribosomal active site, which is still implanted in the modern ribosome. I therefore initiated an experimental study aimed at revealing RNA chains dimerize for acquiring a conformation resembling the PTC and that can bind substrates and catalyze the formation of peptide bonds. Herein I discuss biochemical evidence supporting the existence of an ancestral dimeric proto-ribosome, as well as of compounds that could serve as its substrates. These include the development of structural tools for approaching one of the key questions in origin of life, namely: did the ancient translation apparatus survive selection pressure and are its vestiges are embedded within the modern ribosome?

References

1. Marshall RA, Aitken CE, Dorywalska M, Puglisi JD (2008) Translation at the single-molecule level. Annu Rev Biochem 77:177–203
2. Ramakrishnan V (2008) What we have learned from ribosome structures. Biochem Soc Trans 36(Pt 4):567–574
3. Steitz TA (2008) A structural understanding of the dynamic ribosome machine. Nat Rev Mol Cell Biol 9(3):242–253
4. Yonath A (2009) Large facilities and the evolving ribosome, the cellular machine for genetic-code translation. J R Soc Interface 6(Suppl 5):S575–S585
5. Yonath A, Bashan A (2004) Ribosomal crystallography: initiation, peptide bond formation, and amino acid polymerization are hampered by antibiotics. Annu Rev Microbiol 58:233–251
6. Yonath A, Muessig J, Tesche B, Lorenz S, Erdmann VA, Wittmann HG (1980) Crystallization of the large ribosomal subunit from *B. stearothermophilus*. Biochem Int 1:315–428
7. Schluenzen F, Tocilj A, Zarivach R, Harms J, Gluehmann M, Janell D, Bashan A, Bartels H, Agmon I, Franceschi F, Yonath A (2000) Structure of functionally activated small ribosomal subunit at 3.3 angstroms resolution. Cell 102(5):615–623
8. Wimberly BT, Brodersen DE, Clemons WM Jr, Morgan-Warren RJ, Carter AP, Vonrhein C, Hartsch T, Ramakrishnan V (2000) Structure of the 30S ribosomal subunit. Nature 407(6802):327–339
9. Ban N, Nissen P, Hansen J, Moore PB, Steitz TA (2000) The complete atomic structure of the large ribosomal subunit at 2.4 A resolution. Science 289(5481):905–920
10. Harms J, Schluenzen F, Zarivach R, Bashan A, Gat S, Agmon I, Bartels H, Franceschi F, Yonath A (2001) High resolution structure of the large ribosomal subunit from a mesophilic eubacterium. Cell 107(5):679–688
11. Schuwirth BS, Borovinskaya MA, Hau CW, Zhang W, Vila-Sanjurjo A, Holton JM, Cate JHD (2005) Structures of the bacterial ribosome at 3.5 A resolution. Science 310(5749):827–834
12. Korostelev A, Trakhanov S, Laurberg M, Noller HF (2006) Crystal structure of a 70S ribosome-tRNA complex reveals functional interactions and rearrangements. Cell 126:1065–1077
13. Selmer M, Dunham CM, FVt Murphy, Weixlbaumer A, Petry S, Kelley AC, Weir JR, Ramakrishnan V (2006) Structure of the 70S ribosome complexed with mRNA and tRNA. Science 313(5795):1935–1942
14. Nissen P, Kjeldgaard M, Nyborg J (2000) Macromolecular mimicry. EMBO J 19(4):489–495
15. Agmon I, Bashan A, Zarivach R, Yonath A (2005) Symmetry at the active site of the ribosome: structural and functional implications. Biol Chem 386(9):833–844
16. Bashan A, Agmon I, Zarivach R, Schluenzen F, Harms J, Berisio R, Bartels H, Franceschi F, Auerbach T, Hansen HAS, Kossoy E, Kessler M, Yonath A (2003) Structural basis of the ribosomal machinery for peptide bond formation, translocation, and nascent chain progression. Mol Cell 11:91–102
17. Bashan A, Yonath A (2008) Correlating ribosome function with high-resolution structures. Trends Microbiol 16(7):326–335
18. Bashan A, Zarivach R, Schluenzen F, Agmon I, Harms J, Auerbach T, Baram D, Berisio R, Bartels H, Hansen HA, Fucini P, Wilson D, Peretz M, Kessler M, Yonath A (2003)

Ribosomal crystallography: peptide bond formation and its inhibition. Biopolymers 70(1):19–41

19. Gregory ST, Dahlberg AE (2004) Peptide bond formation is all about proximity. Nat Struct Mol Biol 11:586–587

20. Wekselman I, Davidovich C, Agmon I, Zimmerman E, Rozenberg H, Bashan A, Berisio R, Yonath A (2009) Ribosome's mode of function: myths, facts and recent results. J Pept Sci 15(3):122–130

21. Zimmerman E, Yonath A (2009) Biological implications of the ribosome's stunning stereochemistry. ChemBioChem 10(1):63–72

22. Carter AP, Clemons WM Jr, Brodersen DE, Morgan-Warren RJ, Hartsch T, Wimberly BT, Ramakrishnan V (2001) Crystal structure of an initiation factor bound to the 30S ribosomal subunit. Science 291(5503):498–501

23. Pioletti M, Schluenzen F, Harms J, Zarivach R, Gluehmann M, Avila H, Bashan A, Bartels H, Auerbach T, Jacobi C, Hartsch T, Yonath A, Franceschi F (2001) Crystal structures of complexes of the small ribosomal subunit with tetracycline, edeine and IF3. EMBO J 20(8):1829–1839

24. Auerbach T, Mermershtain I, Bashan A, Davidovich C, Rosenberg H, Sherman DH, Yonath A (2009) Structural basis for the antibacterial activity of the 12-membered-ring mono-sugar macrolide methymycin. Biotechnolog 84:24–35

25. Blaha G, Gurel G, Schroeder SJ, Moore PB, Steitz TA (2008) Mutations outside the anisomycin-binding site can make ribosomes drug-resistant. J Mol Biol 379(3):505–519

26. Hansen JL, Moore PB, Steitz TA (2003) Structures of five antibiotics bound at the peptidyl transferase center of the large ribosomal subunit. J Mol Biol 330(5):1061–1075

27. Harms J, Schluenzen F, Fucini P, Bartels H, Yonath A (2004) Alterations at the peptidyl transferase centre of the ribosome induced by the synergistic action of the streptogramins dalfopristin and quinupristin. BMC Biol 2(1):4;1–10

28. Ippolito JA, Kanyo ZF, Wang D, Franceschi FJ, Moore PB, Steitz TA, Duffy EM (2008) Crystal structure of the oxazolidinone antibiotic linezolid bound to the 50S ribosomal subunit. J Med Chem 51(12):3353–3356

29. Schluenzen F, Zarivach R, Harms J, Bashan A, Tocilj A, Albrecht R, Yonath A, Franceschi F (2001) Structural basis for the interaction of antibiotics with the peptidyl transferase centre in eubacteria. Nature 413(6858):814–821

30. Amit M, Berisio R, Baram D, Harms J, Bashan A, Yonath A (2005) A crevice adjoining the ribosome tunnel: hints for cotranslational folding. FEBS Lett 579(15):3207–3213

31. Berisio R, Harms J, Schluenzen F, Zarivach R, Hansen HA, Fucini P, Yonath A (2003) Structural insight into the antibiotic action of telithromycin against resistant mutants. J Bacteriol 185(14):4276–4279

32. Hansen JL, Ippolito JA, Ban N, Nissen P, Moore PB, Steitz TA (2002) The structures of four macrolide antibiotics bound to the large ribosomal subunit. Mol Cell 10(1):117–128

33. Schluenzen F, Harms JM, Franceschi F, Hansen HA, Bartels H, Zarivach R, Yonath A (2003) Structural basis for the antibiotic activity of ketolides and azalides. Structure (Camb) 11(3):329–338

34. Tu D, Blaha G, Moore PB, Steitz TA (2005) Structures of MLSBK antibiotics bound to mutated large ribosomal subunits provide a structural explanation for resistance. Cell 121:257–270

35. Voss NR, Gerstein M, Steitz TA, Moore PB (2006) The geometry of the ribosomal polypeptide exit tunnel. J Mol Biol 360(4):893–906

36. Agmon I, Amit M, Auerbach T, Bashan A, Baram D, Bartels H, Berisio R, Greenberg I, Harms J, Hansen HA, Kessler M, Pyetan E, Schluenzen F, Sittner A, Yonath A, Zarivach R (2004) Ribosomal crystallography: a flexible nucleotide anchoring tRNA translocation, facilitates peptide-bond formation, chirality discrimination and antibiotics synergism. FEBS Lett 567(1):20–26

37. Cannone JJ, Subramanian S, Schnare MN, Collett JR, D'Souza LM, Du Y, Feng B, Lin N, Madabusi LV, Müller KM, Pande N, Shang Z, Yu N, Gutell RR (2002) The Comparative

RNA Web (CRW) site: an online database of comparative sequence and structure information for ribosomal, intron, and other RNAs. BMC Bioinformatics 3(2):1–31

38. Douthwaite S, Aagaard C (1993) Erythromycin binding is reduced in ribosomes with conformational alterations in the 23 S rRNA peptidyl transferase loop. J Mol Biol 232(3): 725–731

39. Kavanagh F, Hervey A, Robbins WJ (1951) Antibiotic substances from basidiomycetes. VIII. Pleurotus mutilus (Fr.) Sacc. & Pleurotus passeckerianus Pilat. Proc Natl Acad Sci U S A 37:570–574

40. Egger H, Reinshagen H (1976) New pleuromutilin derivatives with enhanced antimicrobial activity. II. Structure–activity correlations. J Antibiot (Tokyo) 29(9):923–927

41. Hildebrandt J, Berner H, Laber G, Schuetze E, Turnowsky F (1983) TDM 85.530—a pleuromutilin derivative and its spectrum of antibacterial activity in vitro. In: Proceedings of the 13th international congress of chemotherapy, vol 5, pp 108/124–108/128

42. Hoegenauer G, Brunowsky W (1983) Mode of action of TDM 85-530. In: Proceedings of the 13th international congress of chemotherapy, vol 5, pp 108/133–108/137

43. Von Graevenitz A, Bucher C (1983) In vitro activity of TDM 85'530 against selected aerobic bacteria. In: Proceedings of the 13th international congress of chemotherapy, vol 5, pp 108/103–108/105

44. Schuster I, Fleschurz C, Hildebrandt J, Turnowsky F, Zsutty H, Kretschmer G, Spitzy KH, Karrer K (1983) Binding and degradation of TDM 85-530 by a microsomal Cyt P-450 form from man, rat and mouse in vitro. In: Proceedings of the 13th international congress of chemotherapy, vol 5, pp 108/142–108/146

45. Ulrich G, Stephen A, Obenaus H, Baumgartner R, Walzl H, Brueggemann S, Schmid B, Racine R, Schatz F, Haberl H, Spitzy KH, Karrer K (1983) TDM 85-530: toxicity in laboratory animals, tolerance and pharmacokinetics after oral application to volunteers. In: Proceedings of the 13th international congress of chemotherapy, vol 5, pp 108/153–108/157

46. Champney WS, Rodgers WK (2007) Retapamulin inhibition of translation and 50S ribosomal subunit formation in Staphylococcus aureus cells. Antimicrob Agents Chemother 51(9):3385–3387

47. Goldstein EJ, Citron DM, Merriam CV, Warren YA, Tyrrell KL, Fernandez HT (2006) Comparative in vitro activities of retapamulin (SB-275833) against 141 clinical isolates of Propionibacterium spp., including 117 P. acnes isolates. Antimicrob Agents Chemother 50(1):379–381

48. Jones RN, Fritsche TR, Sader HS, Ross JE (2006) Activity of retapamulin (SB-275833), a novel pleuromutilin, against selected resistant gram-positive cocci. Antimicrob Agents Chemother 50(7):2583–2586

49. Kosowska-Shick K, Clark C, Credito K, McGhee P, Dewasse B, Bogdanovich T, Appelbaum PC (2006) Single- and multistep resistance selection studies on the activity of retapamulin compared to other agents against Staphylococcus aureus and Streptococcus pyogenes. Antimicrob Agents Chemother 50(2):765–769

50. Pankuch GA, Lin G, Hoellman DB, Good CE, Jacobs MR, Appelbaum PC (2006) Activity of retapamulin against Streptococcus pyogenes and Staphylococcus aureus evaluated by agar dilution, microdilution, E-test, and disk diffusion methodologies. Antimicrob Agents Chemother 50(5):1727–1730

51. Ross JE, Jones RN (2005) Quality control guidelines for susceptibility testing of retapamulin (SB-275833) by reference and standardized methods. J Clin Microbiol 43(12):6212–6213

52. Yan K, Madden L, Choudhry AE, Voigt CS, Copeland RA, Gontarek RR (2006) Biochemical characterization of the interactions of the novel pleuromutilin derivative retapamulin with bacterial ribosomes. Antimicrob Agents Chemother 50(11):3875–3881

53. Scangarella-Oman NE, Shawar RM, Bouchillon S, Hoban D (2009) Microbiological profile of a new topical antibacterial: retapamulin ointment 1%. Expert Rev Anti Infect Ther 7(3):269–279

54. Shawar R, Scangarella-Oman N, Dalessandro M, Breton J, Twynholm M, Li G, Garges H (2009) Topical retapamulin in the management of infected traumatic skin lesions. Ther Clin Risk Manag 5(1):41–49

55. Yang LP, Keam SJ (2008) Retapamulin: a review of its use in the management of impetigo and other uncomplicated superficial skin infections. Drugs 68(6):855–873

56. Yang LP, Keam SJ (2008) Spotlight on retapamulin in impetigo and other uncomplicated superficial skin infections. Am J Clin Dermatol 9(6):411–413

57. Hunt E (2000) Pleuromutilin antibiotics. Drugs Future 25(11):1163–1168

58. Brooks G, Burgess W, Colthurst D, Hinks JD, Hunt E, Pearson MJ, Shea B, Takle AK, Wilson JM, Woodnutt G (2001) Pleuromutilins. Part 1. The identification of novel mutilin 14-carbamates. Bioorg Med Chem 9(5):1221–1231

59. Hodgin LA, Hogenauer G (1974) The mode of action of pleuromutilin derivatives. Effect on cell-free polypeptide synthesis. Eur J Biochem 47(3):527–533

60. Schluenzen F, Pyetan E, Fucini P, Yonath A, Harms J (2004) Inhibition of peptide bond formation by pleuromutilins: the structure of the 50S ribosomal subunit from *Deinococcus radiodurans* in complex with tiamulin. Mol Microbiol 54(5):1287–1294

61. Hogenauer G (1975) The mode of action of pleuromutilin derivatives. Location and properties of the pleuromutilin binding site on *Escherichia coli* ribosomes. Eur J Biochem 52(1):93–98

62. Dornhelm P, Hogenauer G (1978) The effects of tiamulin, a semisynthetic pleuromutilin derivative, on bacterial polypeptide chain initiation. Eur J Biochem 91(2):465–473

63. Davidovich C, Bashan A, Auerbach-Nevo T, Yaggie RD, Gontarek RR, Yonath A (2007) Induced-fit tightens pleuromutilins binding to ribosomes and remote interactions enable their selectivity. Proc Natl Acad Sci U S A 104(11):4291–4296

64. Kehrenberg C, Schwarz S, Jacobsen L, Hansen LH, Vester B (2005) A new mechanism for chloramphenicol, florfenicol and clindamycin resistance: methylation of 23S ribosomal RNA at A2503. Mol Microbiol 57(4):1064–1073

65. Long KS, Poehlsgaard J, Kehrenberg C, Schwarz S, Vester B (2006) The Cfr rRNA methyltransferase confers resistance to phenicols, lincosamides, oxazolidinones, pleuromutilins, and streptogramin A antibiotics. Antimicrob Agents Chemother 50(7): 2500–2505

66. Miller K, Dunsmore CJ, Fishwick CW, Chopra I (2008) Linezolid and tiamulin cross-resistance in *Staphylococcus aureus* mediated by point mutations in the peptidyl-transferase center. Antimicrob Agents Chemother 52(5):1737–1742

67. Pringle M, Poehlsgaard J, Vester B, Long KS (2004) Mutations in ribosomal protein L3 and 23S ribosomal RNA at the peptidyl transferase centre are associated with reduced susceptibility to tiamulin in Brachyspira spp. isolates. Mol Microbiol 54(5):1295–1306

68. Smith LK, Mankin AS (2008) Transcriptional and translational control of the mlr operon which confers resistance to seven classes of protein synthesis inhibitors. Antimicrob Agents Chemother 52(5):1703–1712

69. Toh SM, Xiong L, Arias CA, Villegas MV, Lolans K, Quinn J, Mankin AS (2007) Acquisition of a natural resistance gene renders a clinical strain of methicillin-resistant *Staphylococcus aureus* resistant to the synthetic antibiotic linezolid. Mol Microbiol 64(6):1506–1514

70. Weller J, Hill WE (1991) Probing the initiation complex formation on *E. coli* ribosomes using short complementary DNA oligomers. Biochimie 73(7–8):971–981

71. Tapprich WE, Hill WE (1986) Involvement of bases 787–795 of *Escherichia coli* 16S ribosomal RNA in ribosomal subunit association. Proc Natl Acad Sci U S A 83(3):556–560

72. Hill WE, Tassanakajohn A, Tapprich WE (1990) Interaction of tRNA with domain II of 23S rRNA. Biochim Biophys Acta 1050(1–3):45–50

73. Marconi RT, Hill WE (1989) Evidence for a tRNA/rRNA interaction site within the peptidyltransferase center of the *Escherichia coli* ribosome. Biochemistry 28(2):893–899

74. Lasater LS, Montesano-Roditis L, Cann PA, Glitz DG (1990) Localization of an oligodeoxynucleotide complementing 16S ribosomal RNA residues 520–531 on the small

subunit of *Escherichia coli* ribosomes: electron microscopy of ribosome–cDNA–antibody complexes. Nucleic Acids Res 18(3):477–485

75. Muralikrishna P, Cooperman BS (1991) A photolabile oligodeoxyribonucleotide probe of the peptidyltransferase center: identification of neighboring ribosomal components. Biochemistry 30(22):5421–5428

76. Marconi RT, Lodmell JS, Hill WE (1990) Identification of a rRNA/chloramphenicol interaction site within the peptidyltransferase center of the 50 S subunit of the *Escherichia coli* ribosome. J Biol Chem 265(14):7894–7899

77. Shine J, Dalgarno L (1974) The 3′-terminal sequence of *Escherichia coli* 16S ribosomal RNA: complementarity to nonsense triplets and ribosome binding sites. Proc Natl Acad Sci U S A 71(4):1342–1346

78. Jayaraman K, McParland K, Miller P, Ts'o PO (1981) Selective inhibition of *Escherichia coli* protein synthesis and growth by nonionic oligonucleotides complementary to the 3′ end of 16S rRNA. Proc Natl Acad Sci U S A 78(3):1537–1541

79. Murray RW, Melchior EP, Hagadorn JC, Marotti KR (2001) *Staphylococcus aureus* cell extract transcription–translation assay: firefly luciferase reporter system for evaluating protein translation inhibitors. Antimicrob Agents Chemother 45(6):1900–1904

80. Hausner TP, Atmadja J, Nierhaus KH (1987) Evidence that the G2661 region of 23S rRNA is located at the ribosomal binding sites of both elongation factors. Biochimie 69(9): 911–923

81. Saxena SK, Ackerman EJ (1990) Microinjected oligonucleotides complementary to the alpha-sarcin loop of 28 S RNA abolish protein synthesis in Xenopus oocytes. J Biol Chem 265(6):3263–3269

82. Meyer HA, Triana-Alonso F, Spahn CM, Twardowski T, Sobkiewicz A, Nierhaus KH (1996) Effects of antisense DNA against the alpha-sarcin stem-loop structure of the ribosomal 23S rRNA. Nucleic Acids Res 24(20):3996–4002

83. Abelian A, Walsh AP, Lentzen G, Aboul-Ela F, Gait MJ (2004) Targeting the A site RNA of the *Escherichia coli* ribosomal 30 S subunit by 2′-O-methyl oligoribonucleotides: a quantitative equilibrium dialysis binding assay and differential effects of aminoglycoside antibiotics. Biochem J 383(Pt 2):201–208

84. Ding Y, Lawrence CE (2001) Statistical prediction of single-stranded regions in RNA secondary structure and application to predicting effective antisense target sites and beyond. Nucleic Acids Res 29(5):1034–1046

85. Mathews DH, Sabina J, Zuker M, Turner DH (1999) Expanded sequence dependence of thermodynamic parameters improves prediction of RNA secondary structure. J Mol Biol 288(5):911–940

86. Muckstein U, Tafer H, Hackermuller J, Bernhart SH, Stadler PF, Hofacker IL (2006) Thermodynamics of RNA–RNA binding. Bioinformatics 22(10):1177–1182

87. Stark A, Brennecke J, Russell RB, Cohen SM (2003) Identification of Drosophila MicroRNA targets. PLoS Biol 1(3):E60

88. Stull RA, Taylor LA, Szoka FC Jr (1992) Predicting antisense oligonucleotide inhibitory efficacy: a computational approach using histograms and thermodynamic indices. Nucleic Acids Res 20(13):3501–3508

89. Bretscher MS, Marcker KA (1966) Polypeptidyl-sigma-ribonucleic acid and amino-acyl-sigma-ribonucleic acid binding sites on ribosomes. Nature 211(5047):380–384

90. Monro RE, Marcker KA (1967) Ribosome-catalysed reaction of puromycin with a formylmethionine-containing oligonucleotide. J Mol Biol 25(2):347–350

91. Schulze H, Nierhaus KH (1982) Minimal set of ribosomal components for reconstitution of the peptidyltransferase activity. EMBO J 1(5):609–613

92. Noller HF, Hoffarth V, Zimniak L (1992) Unusual resistance of peptidyl transferase to protein extraction procedures. Science 256(5062):1416–1419

93. Khaitovich P, Tenson T, Kloss P, Mankin AS (1999) Reconstitution of functionally active *Thermus aquaticus* large ribosomal subunits with in vitro-transcribed rRNA. Biochemistry 38(6):1780–1788

94. Khaitovich P, Mankin AS, Green R, Lancaster L, Noller HF (1999) Characterization of functionally active subribosomal particles from *Thermus aquaticus*. Proc Natl Acad Sci U S A 96(1):85–90
95. Anderson RM, Kwon M, Strobel SA (2007) Toward ribosomal RNA catalytic activity in the absence of protein. J Mol Evol 64(4):472–483
96. Nitta I, Kamada Y, Noda H, Ueda T, Watanabe K (1998) Reconstitution of peptide bond formation with *Escherichia coli* 23S ribosomal RNA domains. Science 281(5377):666–669
97. Nitta I, Kamada Y, Noda H, Ueda T, Watanabe K (1999) Peptide bond formation: retraction. Science 283(5410):2019–2020
98. Nitta I, Ueda T, Watanabe K (1998) Possible involvement of *Escherichia coli* 23S ribosomal RNA in peptide bond formation. RNA 4(3):257–267
99. Nitta I, Ueda T, Watanabe K (1999) Retraction. RNA 5(5):707
100. Agmon I, Bashan A, Yonath A (2006) On ribosome conservation and evolution. Isr J Ecol Evol 52:359–379
101. Agmon I, Davidovich C, Bashan A, Yonath A (2009) Identification of the prebiotic translation apparatus within the contemporary ribosome. http://precedings.nature.com/documents/2921/version/1
102. Sato NS, Hirabayashi N, Agmon I, Yonath A, Suzuki T (2006) Comprehensive genetic selection revealed essential bases in the peptidyl-transferase center. Proc Natl Acad Sci U S A 103(42):15386–15391
103. Bieling P, Beringer M, Adio S, Rodnina MV (2006) Peptide bond formation does not involve acid-base catalysis by ribosomal residues. Nat Struct Mol Biol 13(5):424–428
104. Moore PB, Steitz TA (2003) The structural basis of large ribosomal subunit function. Annu Rev Biochem 72:813–850
105. Weinger JS, Strobel SA (2007) Exploring the mechanism of protein synthesis with modified substrates and novel intermediate mimics. Blood Cells Mol Dis 38(2):110–116
106. Weinger JS, Parnell KM, Dorner S, Green R, Strobel SA (2004) Substrate-assisted catalysis of peptide bond formation by the ribosome. Nat Struct Mol Biol 11(11):1101–1106
107. Yonath A (2003) Ribosomal tolerance and peptide bond formation. Biol Chem 384(10):1411–1419
108. Youngman EM, Brunelle JL, Kochaniak AB, Green R (2004) The active site of the ribosome is composed of two layers of conserved nucleotides with distinct roles in peptide bond formation and peptide release. Cell 117(5):589–599

Chapter 2
Methods

2.1 Structural Study of Pleuromutilin Antibiotics

2.1.1 Crystallization and Data Collection

D50S crystals were obtained as previously described [1]. Individual crystals were soaked in harvesting solution at room temperature with 0.01 mM retapamulin (SB-275833) or SB-280080 for 12 h, or 0.1 mM SB-571519 for 8 h and than transferred into an anti-freeze solution [1] for 10 min in the presence of the pleuromutilin antibiotics at adequate concentrations. This was followed by flash cooling in liquid nitrogen. X-ray diffraction data were collected at 85 K. A few crystals were needed for yielding complete datasets of D50S complexes with SB-275833 and SB-280080. Importantly, a complete set with significant redundancy could be collected from a single crystal of the SB-571519 complex (Table 2.1).

2.1.2 Data Processing, Structure Solution and Refinement

Data were processed with HKL2000 [2] and CCP4 package suite [3]. The pleuromutilin derivatives binding sites were unambiguously located in a sigmaa-weighted difference electron density maps using the apo structure of D50S (PDB ID code 1NKW) as reference, after refinement by CNS [4]. Because no crystal structure of any isolated pleuromutilin derivative is available, initial models were generated from tiamulin coordinates (PDB ID code 1XBP) modified by using insight II (Accelrys, San Diego, CA) and CORINA (www.molecular-networks.com/software/category/gen3dcoord.html), followed by energy minimization with no X-ray terms. The maps were traced interactively using O [5] and COOT [6], followed by subsequent restraint CNS minimization for the entire complex. The same randomly selected 5% of the data were omitted for cross validation of all refinement

C. Davidovich, *Targeting Functional Centers of the Ribosome*, Springer Theses, DOI: 10.1007/978-3-642-16931-1_2, © Springer-Verlag Berlin Heidelberg 2011

Table 2.1 Data statistics

Compound	SB-571519	SB-280080	Retapamulin (SB-275833)
Crystal information			
Space group	I222	I222	I222
a (Å)	170.4	170.5	170.1
b (Å)	405.8	412.7	405.9
c (Å)	703.8	696.7	695.2
Diffraction data statistics			
X-ray source	ID19, SBC/APS	ID19, SBC/APS	ID19, SBC/APS
Wavelength (Å)	1.033201	0.97933757	0.979290
Number of crystals	1	7	9
Crystal oscillation (°)	0.3–0.8	0.3–0.4	0.3–0.5
Resolution (Å)	30–3.50 (3.62–3.50)	30–3.56 (3.69–3.56)	30–3.66 (3.79–3.66)
Unique reflections	284,085	263,433	243,598
Observed reflections	1,400,909	1,541,408	1,973,738
Redundancy	4.9 (4.1)	5.9 (3.2)	8.1 (3.7)
Completeness (%)	92.7 (84.9)	90.7 (70.2)	93.1 (70.0)
$<I>/<\sigma>$	6.3 (1.5)	7.2 (1.5)	9.1 (1.4)
R-merge (%)	18.4 (79.7)	16.8 (70.2)	19.0 (79.1)
Refinement			
R-factor (%)	27.5	27.6	26.0
R-free (5%) (%)	33.4	33.8	33.4
rmsd bonds (Å)	0.008	0.008	0.008
rmsd angles (°)	1.4	1.4	1.4

procedures. The resulting coordinates have been deposited at the Protein Data Bank with PDB ID codes 2OGM, 2OGN, and 2OGO.

2.2 Comparative Structural Analysis to Reveal the Structural Basis for Cross-Resistance Between PTC Antibiotics

To perform general structural and statistical analysis, we generated a database holding information on the nucleotides involved in resistance or reduced susceptibility to PTC antibiotics (Table 2.2). This database includes information for each nucleotide, the antibiotics families affected by it, and the distance spans between this nucleotide to the most proximal atom of the bound antibiotic. Precisely, this distance represented the minimum value within a set of distances spans between all possible pairs of two atoms (excluding hydrogens); one in the relevant nucleotide and the other within the corresponding antibiotic molecule. Structural analysis was based on crystal structures of complexes of D50S with various antibiotics, including the phenicol chloramphenicol, the lincosamide clindamycin, the streptogramin$_A$ dalfopristin, the pleuromutilin tiamulin, and the oxazolidinone linezolid. Crystal structures of D50S with other pleuromutilins, methymycin, and lankacidin were not included as currently no resistance data are available for them.

Biochemical data on antibiotic resistance in the archaeon *H. halobium*, and crystallographic data obtained for complexes of PTC antibiotics with the large ribosomal subunit from the archaeon *H. marismortui*, were also included. As these archaeons may resemble eukaryotes and since known pathogens are typically

Table 2.2 Resistance determinants in the PTC

Nucleotide	Appeared clinically [Y/N]	Antibiotic (family)	Bacterial strain	References
2032	N	Chloramphenicol (phenicol)	*E. coli*	[22]
			B. hyodysenteriae	[23]
		Clindamycin (lincosamide)	*E. coli*	[22]
		Linezolid (oxazolidinine)	*E. coli*	[24]
		Tiamulin (pleuromutilin)	*B. hyodysenteriae*	[23]
2055	N	Tiamulin (pleuromutilin)	*B. pilosicoli*	[23]
2057	N	Chloramphenicol (phenicol)	*E. coli*	[22, 25]
2058	Y	Chloramphenicol (phenicol)	*E. coli*	[22]
		Clindamycin (lincosamide)	*E. coli*	[22]
2059	N	Virginiamycin M1 and Pristinamycin II (streptogramins A)	*H. halobium*	[26]
2062	Y	Chloramphenicol (phenicol)	*H. halobium*	[27]
		Linezolid (oxazolidinone)	*H. halobium*	[28]
		Pristinamycin II (streptogramins A)	*S. pneumoniae*	[29]
		Pristinamycin I together with pristinamycin II or dalfopristin together with quinupristin (streptogramins A and B)	*S. pneumoniae*	[29]
2447	N	Linezolid (oxazolidinone)	*E. coli*	[24]
			M. smegmatis	[30]
		Tiamulin (pleuromutilin)	*B. hyodysenteriae*	[23]
			E. coli	[31]
		Chloramphenicol (phenicol)	*E. coli* and *B. stearothermophilus*	[32]
2451	N	Chloramphenicol (phenicol)	*E. coli* and *B. stearothermophilus*	[32]
2452	N	Chloramphenicol (phenicol)	*H. halobium*	[27]
		Linezolid (oxazolidinone)	*H. halobium*	[28]
2453	N	Linezolid (oxazolidinone)	*H. halobium*	[28]
2499	N	Linezolid (oxazolidinone)	*H. halobium*	[28]
		Tiamulin (pleuromutilin)	*B. hyodysenteriae*	[23]
2500	Y	Linezolid (oxazolidinone)	*S. aureus* (MRSA)	[33]
			H. halobium	[28]
			E. coli	[35]
		Tiamulin (pleuromutilin)	*E. coli*	[31]

(continued)

Table 2.2 (continued)

Nucleotide	Appeared clinically [Y/N]	Antibiotic (family)	Bacterial strain	References
2503	Y	Chloramphenicol and florfenicol (phenicols)	*E. coli*	[34]
			S. aureus and *E. coli*	[35, 36]
			S. aureus	[37]
		Clindamycin (lincosamide)	*S. aureus* and *E. coli*	[35, 36]
			S. aureus	[37]
		Linezolid (oxazolidinones)	*S. aureus* and *E. coli*	[35, 36]
			S. aureus	[37]
		Tiamulin and Valnemulin (pleuromutilins)	*S. aureus* and *E. coli*	[35, 36]
		virginamycin M$_1$ (streptogramins A) and dalfopristin/quinupristin (streptogramins A and B)	*S. aureus* and *E. coli*	[35, 36]
			H. halobium	[26]
2504	N	Linezolid (oxazolidinone)	*H. halobium*	[28]
		Tiamulin (pleuromutilin)	*Brachyspira pilosicoli*	[23]
			B. hyodysenteriae	[23]
2505	N	Linezolid (oxazolidinone)	*Enterococcus*	[38]
			E. faecalis	[39]
2572	N	Tiamulin (pleuromutilin)	*B. hyodysenteriae*	[23]
2576	Y	Linezolid (oxazolidinone)	*Enterococcus*	[38]
			Enterococcus	[40]
			S. aureus	[41]
			E. faecium	[42]
			S. aureus	[43]
			S. aureus	[44]
			E. faecium	[45]
		Tiamulin (pleuromutilin)	*S. aureus*	[31]

Nucleotides are according to *E. coli* numbering system

eubacteria, distances and data analyses were performed separately for eubacteria and for archaea. Calculations and database analysis were performed by MATLAB (MathWorks, Inc., Natick, MA). In silico mutagenesis and structural analysis were performed by COOT [6] and PyMol (DeLano Scientific, Palo Alto, CA, USA). PyMol was also used for image rendering. PDB ID codes are shown when relevant.

2.3 Antisense Oligonucleotides for Targeting Functional Ribosomal Centers

2.3.1 Database Construction

The 23S rRNA of the large ribosomal subunit of eubacteria was selected as a model for initial study. First, we constructed a database that contained structural and thermodynamical information (MathWorks, Inc., Natick, MA). The parameters that were incorporated into the database are as follows:

1. *Mutation sites*: Includes mutations in nucleotides that were previously related with one of the following categories: (i) antibiotic resistance or antibiotic enhancement, (ii) slow bacterial growth or reduce ribosomal activity and (iii) lethality. A total number of 132 mutations obeying at least one of these categories were selected out of the rRNA mutations database of Kathleen Triman (ribosome.fandm.edu).

2. *Stacking distance*: Including distances between the bases of each two sequent nucleotides (i.e. nucleotides i and i + 1), using N3 as a reference atom. The distances were calculated according to the coordinates of the large ribosomal subunit from *D. Radiodurans* (1NKW). A small $N3_i$ to $N3_{i+1}$ distance is a rough assessment for a stacking interaction between nucleobases, while a long distance indicates poor stacking.

3. *Base pairing*: Indicates Watson–Crick and Wobble pairs, as appeared in the 3D structure, detected by X3DNA [7].

4. *Surface accessibility*: Was calculated for each nucleotide separately; for the entire residue and for the backbone only (phosphate and ribose). Calculation done in the presence of all the large ribosomal subunit components (23S, 5S and r-proteins), using a probe radius of 1.4 Å (Crystallography & NMR System (CNS), [4]).

5. *Thermodynamics parameters*: Done by our collaborator, Joseph Sofaer, from the research group of Tamar Schlick, Faculty of Chemistry, NYU, using the computer program OligoWalk [8]. These parameters were calculated for all the possible 5–25 bases long antisense DNA oligonucleotides for the 23S rRNA of *E. coli*: a total sum of 60,690 theoretical antisense sequences. The following parameters were derived for each of them:

   a. *Target structure energy*: The energy associated with braking the 2D structure of the target rRNA. The 2D structure of the 23S rRNA used to generate this parameter.

   b. *Intramolecular energy*: Calculated for secondary structures that might be formed by the antisense itself.

   c. *Intermolecular energy*: Considered antisense–antisense interactions.

   d. *Duplex formation energy*: The energy gain of binding an antisense to the target sequence.

   e. *Total binding energy*: Incorporating all of the thermodynamic parameters mentioned above to yield the overall binding energy.

   f. *Tm of the duplex*: An estimate for the temperature by which 50% of the target sequences are bound to antisense oligonucleotides. This calculation neglected the structures of the target and of the antisense oligonucleotide, and assumed an excess of the antisense over the target. Tm calculated for antisense concentration of 1 μM.

2.3.2 In Vitro Transcription–Translation System for Ribosome Activity Assay

In order to characterize IC_{50} values for the selected antisense oligonucleotides we constructed an in vitro transcription–translation system of *E. coli*, with firefly luciferase as a reporter gene. The procedure for in vitro transcription–translation system was used as previously described [9, 10], with some modifications.

2.3.2.1 *E. coli* Cell Extract Preparation

E. coli BL21(DE3) strain (Stratagene) fermented in LB medium (250 rpm at 37°C). Induction with 1 mM IPTG, for expression of T7 RNA polymerase, performed at 0.5 OD_{600}, followed by additional 4 h of incubation (250 rpm at 37°C) before harvesting (15 min under 4000 RCF at 4°C). Next, 20 g of bacterial paste were washed twice in 1 L and 500 ml buffer A (60 mM AcOK, 10 mM Tris–OAc pH 8.2, 14 mM $Mg(OAc)_2$, 155 μg/ml DTT, 0.5 μl/ml β-mercaptoethanol), followed by centrifugation (SPA-3000 rotor, 6000 rpm, 15 min, 4°C). The pellet was resuspended with 20 ml of buffer B (60 mM AcOK, 10 mM Tris–OAc pH 8.2, 14 mM $Mg(OAc)_2$, 155 μg/ml DTT). Cells were lysed using a French Press (American Instrument Company, Silver Spring, MA) at 8000 psi. The bacterial lysate was centrifuged to remove cell debris (SS-34 rotor, 11,600 rpm, 30 min, 4°C). The supernatant was re-centrifuged under the same conditions, and then dialyzed with SnakeSkin® 3.5 kDa molecular weight cut-off (MWCO) regenerated cellulose membrane (Pierce Biotechnology, Rockford, IL) against buffer B (750 ml × 4 times, 1–1.5 h each, 4°C). The solution was centrifuged (SS-34 rotor, 5,800 rpm, 10 min, 4°C) followed by incubation of 30 min at 37°C to reduce amounts of endogenous mRNA. Cell extract was divided into 500 μl aliquots, flash-frozen in liquid nitrogen, and stored at −80°C.

2.3.2.2 Transcription–Translation Reaction

Reaction mixtures included: 0.22 v/v *E. coli* cell extract, 1.8 ng/μl pIVEX2.6d plasmid carrying firefly luciferase under T7 promoter (donated by Roy Bar-Ziv, Weizmann Institute of Science), 900 μM of each amino acid (Sigma), 230 μg/ml Creatine Kinase from rabbit muscle (Roche), 150 μg/ml *E. coli* tRNA mixture (Sigma), 50 mM Hepes–KOH (pH 7.5), 9% PEG 8000, 180 mM KGlu, 1.6 mM DTT, 1.1 mM ATP, 0.8 mM GTP, 0.8 mM CTP and 0.8 mM UTP, 0.6 mM cAMP, 70 mM Creatine Phosphate, 30 μg/ml folinic acid, 25 mM NH_4OAc and 16 mM $Mg(OAc)_2$.

30 μl reaction mixtures were incubated for 60 min at 37°C in the presence of different concentrations of the various ODNs. Reactions stopped instantly with the addition of erythromycin solution to final concentration of 5 μM.

2.3.2.3 Luciferase Activity Assay for Translation Quantification

50 µl Luciferase Assay System (Promega) were added to 10 µl of the incubated reaction mixture in a black 96 well plate (Nunc, Roskilde, Denmark), and photoluminescence was instantly measured using SpectraFluor Plus plate reader (Tecan, Maennedorf, Switzerland). Dose response curves were fitted to the experimental data with MATLAB (MathWorks, Inc., Natick, MA) to produce IC_{50} values.

2.3.3 Antisense Oligonucleotides Nomenclature

Antisense oligonucleotides were named according to their length and their first target nucleotide in *E. coli* numbering (e.g. antisense 463-10 is a 10 bases antisense that targets nucleotides 463–472 in *E. coli*).

2.4 Minimal Ribosomal Components with PTC Structure and Function

2.4.1 In Vitro RNA Transcription

DNA constructs were prepared using gene synthesis (purchased from GenScript, NJ) and were obtained in varies scales from few micrograms to few milligrams (Plasmid Mini, Midi, Maxi and Giga Kits, Qiagen). Plasmids were linearized with HgaI or BsaI (New England Biolabs, Madison, WI). RNA transcription and purification curried out from microgram to milligram scales, as described elsewhere [11], using a recombinant His-tagged T7 RNA polymerase (RNAP). T7 RNAP was overexpressed following standard protocols (*E. coli* BL21 strain carrying the gene in pQE8 expression vector was kindly donated by Dr. Irit Sagi and Dr. Barak Akabayov, Weizmann Institute). Mutations in RNA sequences were incorporated either by transcription from custom synthesized single strand DNA templates carrying the desired mutation (Sigma) or were incorporated to plasmids carrying the wild-type sequence using QuikChange Site-Directed Mutagenesis Kit (Stratagene), following manufacturer instructions.

2.4.2 Study of Dimerization Tendency

RNA samples were dissolved in double distilled water and annealed for 1 min at 90°C then flash cooled on ice. Incubation buffer was added to a final concentration of 89 mM tris–borate (TB) pH 7.5 and 15 mM MgCl. Unless otherwise indicated, final RNA concentration in samples was 2 μM. Samples were incubated at 37°C for 30 min and then placed on ice for additional 15 min. For non-denaturing gels 10× glycerol loading dye (50% glycerol, 0.25% w/v bromophenol blue and 0.25% w/v xylene cyanol) was added and the samples were immediately loaded on ice-chilled 12% or 15% polyacrylamide (19:1 polyacrylamide:bisacrylamide) mini gel buffered with 89 mM TB pH 7.5 and 15 mM MgCl. For magnesium-free assays, MgCl$_2$ was replaced with 2 mM EDTA in both samples and gels. Non-denaturing gels were run overnight in icebox, under constant voltage of 40 V (Mini-PRO-TEAN®, BioRad). For denaturing gels 2× formamide loading dye (90% v/v deionized formamide, 0.25% w/v bromophenol blue, 0.25% w/v xylene cyanol, 2 mM EDTA, 100 mM Tris–HCl, pH 7.3) was added to RNA samples following 10 min incubation at 65°C. Next, samples were briefly chilled on ice and loaded on denaturing 12% or 15% polyacrylamide gel with 7 M urea, buffered with 89 mM tris–borate pH 8.3 and 2.5 mM EDTA (TBE). Denaturing gels were run up to 2 h under constant voltage of 150 V (Mini-PROTEAN®, BioRad). Gels were stained with 0.5 μg/ml ethidium bromide (EtBr) and imaged using Bio-Rad Molecular Imager Gel Doc XR system, controlled by the Quantity One software (Bio-Rad).

2.4.3 Electrophoresis Mobility Shift Assay

The studied RNA was dephosphorylated using calf intestinal alkaline phosphatase (New England Biolabs) and 5′-end labeled with ^{33}P (gamma-^{33}P-ATP, Institute of Isotope Co., Ltd., Budapest, Hungary) or ^{32}P (gamma-^{32}P-ATP, PerkinElmer Inc.) using T4 polynucleotide kinase (New England Biolabs) as described elsewhere [11] or by KinaseMax™ Kit (Ambion) according to manufacturer instructions. For homodimer Kd assessments non-labeled RNA of the same sequence in various concentrations was mixed with trace amount of ^{33}P or ^{32}P labeled RNA, and then was subjected to dimerization assay followed by non-denaturing polyacrylamide gel electrophoresis, as described above. For heterodimerization, labeled and unlabeled RNA of different sequences were used. Gels were transferred to Whatman 3 M paper and were dried under a vacuum at 80°C for 2 h using BioRad gel drying apparatus. Dried gels were exposed to a phosphorimager cassette (Molecular Dynamics) for periods ranging from overnight to one week, scanned (Phosphorimager Storm 820, Amersham Pharmacia Biotech) and analyzed (Quantity One software, Bio-Rad).

2.4.4 Size Exclusion Chromatography for the Separation
Between Dimer and Monomer

SEC based separation between the dimer and the monomer was based on previous publications [12–14] with some modifications. Briefly, samples of the studied

RNA were annealed as described above for non-denaturing gels. 10–100 μg RNA samples were loaded on Superdex 200 10/300 GL column (Amersham Pharmacia Biosciences) equilibrated with 20 mM Tris–HCl pH 7.8, 40 mM KCl and 8 mM MgCl$_2$. Load volume was not exceeding 15 μl. Chromatography was performed at 0.5 ml/min at 4–7°C and fractions were collected manually into ice-chilled tubes.

2.4.5 Radiolabeling of Substrates for Peptidyl Transferase Activity Assay

CC-puromycin (CCPmn) or C-puromycin (CPmn) substrates (Dharmacon) were labeled with ^{33}P or ^{32}P as described above for 5′ RNA labeling. 2× formamide loading dye (FD) added to labeled substrates prior to separation from unincorporated radiolabeled ATP by polyacrylamide gel electrophoresis (PAGE), using 12% acrylamide:bis-acrylamide ratio 29:1 gels buffered with 1× TBE and 7 M urea as denaturing agent. Gels run 1.5 h at 80 V (Mini-PROTEAN®, BioRad) in a tank containing 0.5× TBE. Under these conditions CCPmn labeled substrate co-migrated with the xylene cyanol dye. CCPmn and CPmn containing bends were sliced and eluted into Milli-Q water overnight at 4°C and quantified by scintillation counting.

2.4.6 Assay for Peptidyl Transferase Activity

Peptidyl transferase activity for various dimers performed as previously described [15] with some modifications. Briefly, RNA constructs were resuspended in Milli-Q water, annealed 1 min at 90°C and cooled down to room temperature before adding incubation buffer to final concentration of 120 mM KCl, 15 mM MgCl$_2$, 40 mM of either 2-(N-morpholino) ethanesulfonic acid (MES) pH 6.0, TRIS pH 6.8, HEPES pH 7.5 or HEPES pH 8.0. Substrates added to final concentration of 1 mM CCA-phenylalanine-caproic acid-biotin (Dharmacon) and > 2000 cpm ^{33}P or ^{32}P labeled CCPmn or CPmn. Samples contained 2 μM RNA. Control samples were incubated in the absence of RNA and the presence or absence of 1 μM 50S ribosomal subunit from E. coli. To check possible positive contribution of alcohol to peptidyl transferase activity, as was previously observed for similar reactions using larger substrates [16, 17], samples were incubated in the presence or absence of 30% methanol. 10 μl reactions were incubated for periods varying from 2 to 24 h at room temperature, 37 or 50°C. Reactions were quenched with the addition of 18.1 μl of 2× FD following with the addition of 8.1 μl of 5 mg/ml streptavidin (GenScript) aqueous in 150 mM NaCl and 10 mM potassium phosphate, pH 7.2. Samples were loaded on 7% acrylamide:bis-acrylamide 29:1 gels buffered with 1× TBE and 7 M urea as denaturing agent. Gels run 1 h at 80 V in a tank containing 0.5× TBE (Mini-PROTEAN®, BioRad), dried and scanned as described above.

2.4.7 RNA Two-Dimensional Structure Prediction

Two-dimensional structure prediction performed using mFold [18], version 3.2, using the default settings.

2.5 Numbering, Sequence Alignment, and Images

Nucleotides are numbered according to *E. coli* numbering system throughout, unless otherwise mentioned. Multiple sequence alignment was performed by ClustalW [19] and presented by JalView [20]. Figures of three-dimensional structures were generated by Pymol [21].

References

1. Auerbach-Nevo T, Zarivach R, Peretz M, Yonath A (2005) Reproducible growth of well diffracting ribosomal crystals. Acta Crystallogr D Biol Crystallogr 61:713–719
2. Otwinowski Z, Minor W (1997) Processing of X-ray diffraction data collected in oscillation mode, methods in enzymology. In: Carter JCW, Sweet RM (eds) Macromolecular crystallography, vol 276, pp 307–326
3. Bailey S (1994) The CCP4 suite—programs for protein crystallography. Acta Crystallogr D Biol Crystallogr 50:760–763
4. Brunger AT, Adams PD, Clore GM, DeLano WL, Gros P, Grosse-Kunstleve RW, Jiang JS, Kuszewski J, Nilges M, Pannu NS, Read RJ, Rice LM, Simonson T, Warren GL (1998) Crystallography & NMR system: a new software suite for macromolecular structure determination. Acta Crystallogr D Biol Crystallogr 54(Pt 5):905–921
5. Jones TA, Zou JY, Cowan SW, Kjeldgaard M (1991) Improved methods for building protein models in electron density maps and the location of errors in these models. Acta Crystallogr A 47(Pt 2):110–119
6. Emsley P, Cowtan K (2004) Coot: model-building tools for molecular graphics. Acta Crystallogr D Biol Crystallogr 60(Pt 12 Pt 1):2126–2132
7. Lu XJ, Olson WK (2003) 3DNA: a software package for the analysis, rebuilding and visualization of three-dimensional nucleic acid structures. Nucleic Acids Res 31(17):5108–5121
8. Mathews DH, Sabina J, Zuker M, Turner DH (1999) Expanded sequence dependence of thermodynamic parameters improves prediction of RNA secondary structure. J Mol Biol 288(5):911–940
9. Abelian A, Walsh AP, Lentzen G, Aboul-Ela F, Gait MJ (2004) Targeting the A site RNA of the *Escherichia coli* ribosomal 30 S subunit by 2′-O-methyl oligoribonucleotides: a quantitative equilibrium dialysis binding assay and differential effects of aminoglycoside antibiotics. Biochem J 383(Pt 2):201–208
10. Zubay G (1973) In vitro synthesis of protein in microbial systems. Annu Rev Genet 7:267–287
11. Christopher WS (1998) RNA: protein interactions: a practical approach, 1st edn. Oxford University Press, Oxford

12. Kim I, McKenna SA, Viani Puglisi E, Puglisi JD (2007) Rapid purification of RNAs using fast performance liquid chromatography (FPLC). RNA 13(2):289–294

13. Lukavsky PJ, Puglisi JD (2004) Large-scale preparation and purification of polyacrylamide-free RNA oligonucleotides. RNA 10(5):889–893

14. McKenna SA, Kim I, Puglisi EV, Lindhout DA, Aitken CE, Marshall RA, Puglisi JD (2007) Purification and characterization of transcribed RNAs using gel filtration chromatography. Nat Protoc 2(12):3270–3277

15. Schmeing TM, Seila AC, Hansen JL, Freeborn B, Soukup JK, Scaringe SA, Strobel SA, Moore PB, Steitz TA (2002) A pre-translocational intermediate in protein synthesis observed in crystals of enzymatically active 50S subunits. Nat Struct Biol 9(3):225–230

16. Bretscher MS, Marcker KA (1966) Polypeptidyl-sigma-ribonucleic acid and amino-acyl-sigma-ribonucleic acid binding sites on ribosomes. Nature 211(5047):380–384

17. Monro RE, Marcker KA (1967) Ribosome-catalysed reaction of puromycin with a formylmethionine-containing oligonucleotide. J Mol Biol 25(2):347–350

18. Zuker M (1989) On finding all suboptimal foldings of an RNA molecule. Science 244(4900):48–52

19. Thompson JD, Higgins DG, Gibson TJ (1994) CLUSTAL W: improving the sensitivity of progressive multiple sequence alignment through sequence weighting, position-specific gap penalties and weight matrix choice. Nucleic Acids Res 22(22):4673–4680

20. Clamp M, Cuff J, Searle SM, Barton GJ (2004) The Jalview Java alignment editor. Bioinformatics 20(3):426–427

21. DeLano WL (2002) The PyMOL molecular graphics system. DeLano Scientific, San Carlos

22. Douthwaite S (1992) Functional interactions within 23S rRNA involving the peptidyltransferase center. J Bacteriol 174(4):1333–1338

23. Pringle M, Poehlsgaard J, Vester B, Long KS (2004) Mutations in ribosomal protein L3 and 23S ribosomal RNA at the peptidyl transferase centre are associated with reduced susceptibility to tiamulin in Brachyspira spp. isolates. Mol Microbiol 54(5):1295–1306

24. Xiong L, Kloss P, Douthwaite S, Andersen NM, Swaney S, Shinabarger DL, Mankin AS (2000) Oxazolidinone resistance mutations in 23S rRNA of Escherichia coli reveal the central region of domain V as the primary site of drug action. J Bacteriol 182(19):5325–5331

25. Ettayebi M, Prasad SM, Morgan EA (1985) Chloramphenicol-erythromycin resistance mutations in a 23S rRNA gene of Escherichia coli. J Bacteriol 162(2):551–557

26. Porse BT, Garrett RA (1999) Sites of interaction of streptogramin A and B antibiotics in the peptidyl transferase loop of 23 S rRNA and the synergism of their inhibitory mechanisms. J Mol Biol 286(2):375–387

27. Mankin AS, Garrett RA (1991) Chloramphenicol resistance mutations in the single 23S rRNA gene of the archaeon Halobacterium halobium. J Bacteriol 173(11):3559–3563

28. Kloss P, Xiong L, Shinabarger DL, Mankin AS (1999) Resistance mutations in 23 S rRNA identify the site of action of the protein synthesis inhibitor linezolid in the ribosomal peptidyl transferase center. J Mol Biol 294(1):93–101

29. Depardieu F, Courvalin P (2001) Mutation in 23S rRNA responsible for resistance to 16-membered macrolides and streptogramins in Streptococcus pneumoniae. Antimicrob Agents Chemother 45(1):319–323

30. Sander P, Belova L, Kidan YG, Pfister P, Mankin AS, Bottger EC (2002) Ribosomal and non-ribosomal resistance to oxazolidinones: species-specific idiosyncrasy of ribosomal alterations. Mol Microbiol 46(5):1295–1304

31. Miller K, Dunsmore CJ, Fishwick CW, Chopra I (2008) Linezolid and tiamulin cross-resistance in Staphylococcus aureus mediated by point mutations in the peptidyl-transferase center. Antimicrob Agents Chemother 52(5):1737–1742

32. Thompson J, Kim DF, O'Connor M, Lieberman KR, Bayfield MA, Gregory ST, Green R, Noller HF, Dahlberg AE (2001) Analysis of mutations at residues A2451 and G2447 of 23S rRNA in the peptidyltransferase active site of the 50S ribosomal subunit. Proc Natl Acad Sci U S A 98(16):9002–9007

33. Meka VG, Pillai SK, Sakoulas G, Wennersten C, Venkataraman L, DeGirolami PC, Eliopoulos GM, Moellering RC Jr, Gold HS (2004) Linezolid resistance in sequential *Staphylococcus aureus* isolates associated with a T2500A mutation in the 23S rRNA gene and loss of a single copy of rRNA. J Infect Dis 190(2):311–317

34. Vester B, Garrett RA (1988) The importance of highly conserved nucleotides in the binding region of chloramphenicol at the peptidyl transfer centre of *Escherichia coli* 23S ribosomal RNA. EMBO J 7(11):3577–3587

35. Long KS, Poehlsgaard J, Kehrenberg C, Schwarz S, Vester B (2006) The Cfr rRNA methyltransferase confers resistance to phenicols, lincosamides, oxazolidinones, pleuromutilins, and streptogramin a antibiotics. Antimicrob Agents Chemother 50(7):2500–2505

36. Kehrenberg C, Schwarz S, Jacobsen L, Hansen LH, Vester B (2005) A new mechanism for chloramphenicol, florfenicol and clindamycin resistance: methylation of 23S ribosomal RNA at A2503. Mol Microbiol 57(4):1064–1073

37. Toh SM, Xiong L, Arias CA, Villegas MV, Lolans K, Quinn J, Mankin AS (2007) Acquisition of a natural resistance gene renders a clinical strain of methicillin-resistant *Staphylococcus aureus* resistant to the synthetic antibiotic linezolid. Mol Microbiol 64(6):1506–1514

38. Prystowsky J, Siddiqui F, Chosay J, Shinabarger DL, Millichap J, Peterson LR, Noskin GA (2001) Resistance to linezolid: characterization of mutations in rRNA and comparison of their occurrences in vancomycin-resistant enterococci. Antimicrob Agents Chemother 45(7):2154–2156

39. Lobritz M et al (2003) Recombination proficiency influences frequency and locus of mutational resistance to linezolid in *Enterococcus faecalis*. Antimicrob Agents Chemother 47(10):3318–3320

40. Auckland C et al (2002) Linezolid-resistant enterococci: report of the first isolates in the United Kingdom. J Antimicrob Chemother 50(5):743–746

41. Wilson P et al (2003) Linezolid resistance in clinical isolates of *Staphylococcus aureus*. J Antimicrob Chemother 51(1):186–188

42. Rahim S et al (2003) Linezolid-resistant, vancomycin-resistant *Enterococcus faecium* infection in patients without prior exposure to linezolid. Clin Infect Dis 36(11):E146–E148

43. Pillai SK et al (2002) Linezolid resistance in *Staphylococcus aureus*: characterization and stability of resistant phenotype. J Infect Dis 186(11):1603–1607

44. Tsiodras S et al (2001) Linezolid resistance in a clinical isolate of *Staphylococcus aureus*. Lancet 358(9277):207–208

45. Herrero IA, Issa NC, Patel R (2002) Nosocomial spread of linezolid-resistant, vancomycin-resistant Enterococcus faecium. N Engl J Med 346(11):867–869

Chapter 3
Results

3.1 The Structures of D50S/Pleuromutilins Complexes

Complete crystallographic data sets of complexes of D50S with SB-571519, SB-280080, and retapamulin (SB-275833) (Fig. 1.2) yielded electron density maps at 3.50, 3.56, and 3.66 Å resolution, respectively (Table 2.1), in which the location and conformation of each of the three bound compounds were unambiguously resolved (Fig. 3.1). Grouped occupancy refinement yielded a value of approximately 1.0 for all three compounds, confirming that they are quantitatively bound, in accord with the high binding affinity observed (Table 3.1) in a competitive ribosome-binding assay by using a radiolabeled pleuromutilin derivative [1].

The electron density maps of each of the three pleuromutilins complexes (Fig. 3.1a–c) indicate that all of these pleuromutilins are located at the PTC, similar to tiamulin [2], in the vicinity of the location that should have been occupied by the transition state intermediate of peptide bond formation [3]. Their tricyclic cores are almost fully superposed on that observed for tiamulin (Fig. 3.2). All interact with the 23S RNA domain V (Fig. 3.1d–f) by hydrophobic interactions and hydrogen bonds, formed with surrounding nucleotides, namely A2503, U2504, G2505, U2506, C2452, and U2585. The C11 hydroxyl group of all of the compounds is located in a position suitable for hydrogen bonding to G2505 phosphate. In SB-280080 and SB-571519, it can be involved in an additional H-bond with the O2' hydroxyl of A2503 (Fig. 3.1d and f). The additional hydroxyl group of SB-571519 C2 (R_2 in Fig. 1.2) may be involved in polar interactions or an H-bond with O3' or O5' phosphoester of G2505 (Fig. 3.1d), because the distances between its oxygen and G2505 phosphoester oxygens are 3.2 and 3.0 Å, respectively.

As observed for tiamulin [2], the essential [4] C21 keto group (Fig. 1.2) of the C14 extension of all three compounds is involved in two to three hydrogen bonds with G2061. This H-bond network is similar for all three compounds, including the carbamate derivative SB-571519 that utilizes its additional carbonyl as an alternative H-bond acceptor (Fig. 3.1d–f). Apart from these H-bonds with G2061, it

C. Davidovich, *Targeting Functional Centers of the Ribosome*, Springer Theses, DOI: 10.1007/978-3-642-16931-1_3, © Springer-Verlag Berlin Heidelberg 2011

seems that the pleuromutilins' C14 extension is involved only in minor hydrophobic contacts with ribosomal nucleotides.

Among the conformational rearrangements of the rRNA observed upon binding of all three studied pleuromutilins (Fig. 3.1g–i), the most notable is the 40° rotation of U2506 base toward the tricyclic core, a motion that closes tightly the binding pocket on the bound compound. An additional conformational

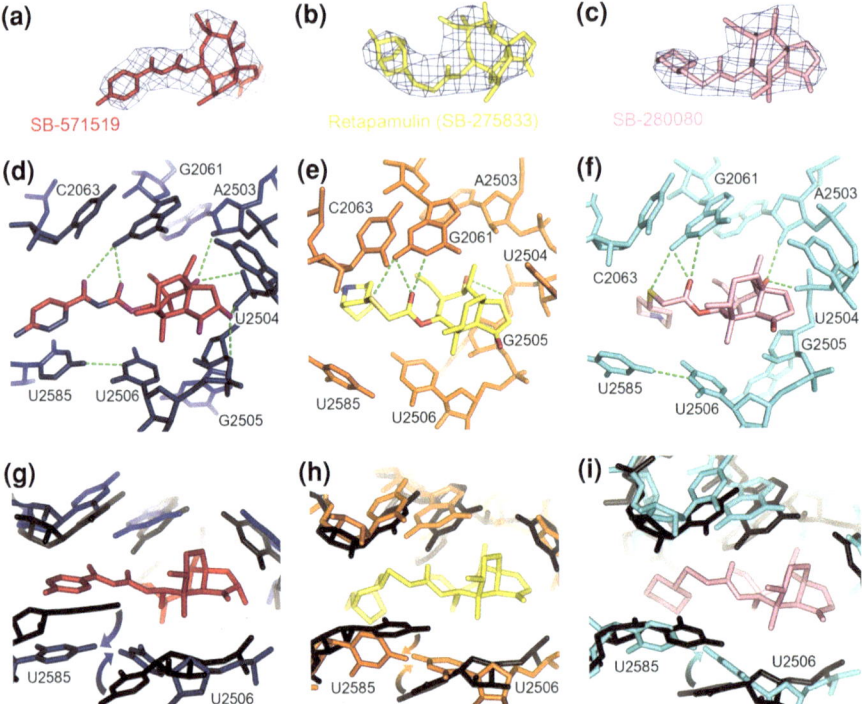

Fig. 3.1 Induced fit binding of pleuromutilin antibiotics to the PTC. Selected regions from the crystal structures of the large ribosomal subunit from *D. radiodurans* complexed with pleuromutilin acyl-carbamate derivative SB-571519 (in *red*: **a**, **d**, and **g**) and two sulfanyl-acetate derivatives retapamulin (SB-275833) (in *yellow*: **b**, **e**, and **h**) and SB-280080 (in *pink*: **c**, **f**, and **i**) at 3.50, 3.66, and 3.56-Å resolution, respectively. **a–c**: 2Fo − Fc electron density maps, contoured at 1.5 σ level. **d–f**: Interactions between pleuromutilin derivatives and 23S rRNA (the 23S rRNA bound conformation is shown in *dark blue* for SB-571519, *orange* for SB-275833, and *cyan* for SB-28008). All show H-bonds between G2061 and pleuromutilins' C14 extension and the possibility of H-bond between C11 hydroxyl and G2505 phosphate. In SB-571519 and SB-280080 complexes, C11 hydroxyl may serve as acceptor in an additional H-bond with the OH group of A2503 O2′, and a nontypical H-Bond is observed between U2585 and U2506. **g–i**: Induced-fit mechanism promotes pleuromutilins binding (color code as in **d–f**). In all, the unbound 23S rRNA (PDB ID code 1NKW) is *black*, and the pleuromutilin-bound conformations are *color*. On binding, U2585 translates away from the C14 extension and U2506 rotates toward the pleuromutilin and thus closes tightly the binding pocket. H-bonds (**d** and **f**) or other interactions (**e**) between the two shifted nucleotides may further stabilize the rRNA conformation at the bound state

Table 3.1 Affinities of pleuromutilin derivatives in binding to *E. coli* ribosomes

Compound	K_d [nM]
SB-275833 (retapamulin)	2.0 ± 0.05
SB-280080	7.5 ± 1.4
SB-571519	11.1 ± 3.2

Values are results $\pm$ standard error from two independent experiments

Fig. 3.2 Four pleuromutilin derivatives superimposed in the binding pocket. **a** A surface representation of the binding pocket. Several nucleotides have been removed to permit a clear view of the binding site. The structure of D50S/SB-517519 was used for surface representation. **b** A side view of pleuromutilins with the 3′ ends of an A-site tRNA mimic and the derived P-site tRNA. All pleuromutilin derivatives presented here are located at the PTC, with their tricyclic core oriented similar to tiamulin and their C14 extensions placed within the PTC

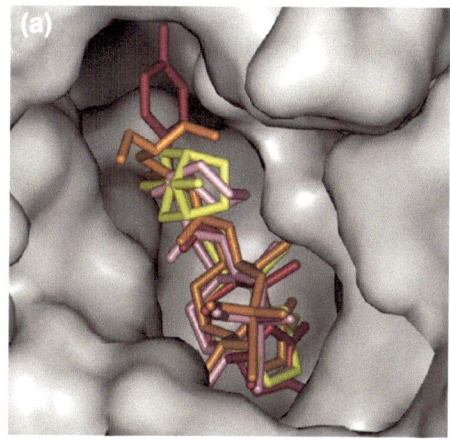

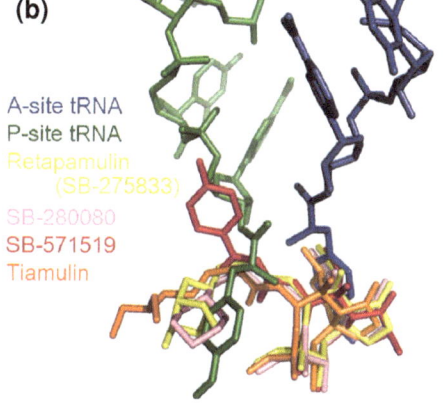

rearrangement, of the flexible base of U2585 [5, 6], was also detected. In eubacteria, in the absence of substrates or inhibitors, this flexible nucleotide [6] is located in a position [7–9] that should interfere or interact with C14 extension of all studied pleuromutilins, including tiamulin. Consequently, to avoid steric hindrance, in the presence of pleuromutilins U2585 undergoes a slight shift, which allows for interactions between its base and that of U2506, thus stabilizing the conformations of both nucleotides in the bound state (Fig. 3.1d–f). In the SB-571519- and SB-280080-bound forms, U2585 forms a single hydrogen bond with

U2506, whereas in the retapamulin complex the distance between the two bases allows for van der Waals or similar interactions.

It was previously suggested that mutations in ribosomal protein L3 reduce bacterial susceptibility to pleuromutilins by indirect influence of the binding pocket conformation [2, 10, 11]. I could unambiguously trace protein L3 Arg-144 (*D. radiodurans* numbering) owing to the high quality of the electron density map of SB-571519/D50S complex, as it was constructed from a complete dataset that was collected form a single crystal with significant redundancy (Table 2.1). This map shows that L3 Arg-144 extends toward the PTC, and interacts electrostatically with U2506 phosphate, but does not make any contacts with the pleuromutilin compound (Fig. 3.3a). In addition, no significant conformational rearrangements

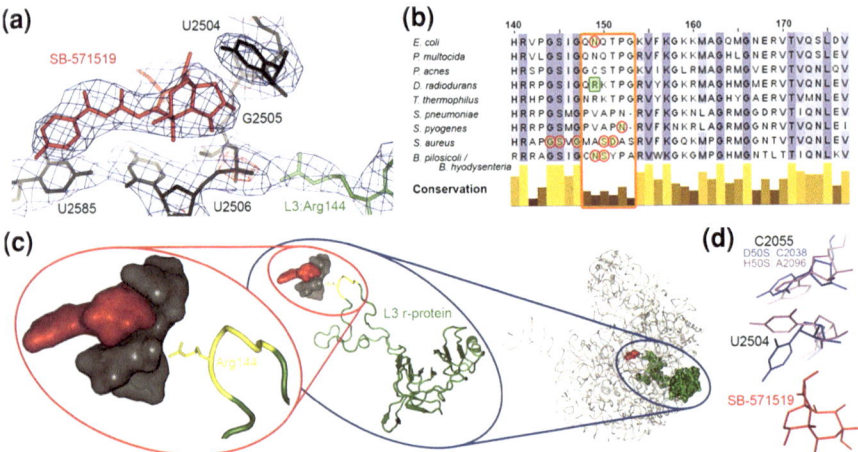

Fig. 3.3 Remote factors acquiring resistance and selectivity. **a–c** Possible contribution of L3 to pleuromutilin resistance. **a** 2 Fo – Fc electron density map of the carbamate-derivative (*red*) SB-571519. 23S rRNA nucleotides are shown in *black*, and protein L3 Arg144 in *green*. *Blue* mesh is contoured at 1.0 σ. The *red* meshes (contoured at 5.0 σ) indicate phosphate locations. **b** Multiple sequence alignment of protein L3 from selected bacteria. *Top numbering* according to *E. coli* (where Arg144 of *D. radiodurans* is Asn149), *bottom bars* indicate conservation among these strains, and *yellow circles* indicate mutations with reduced pleuromutilin susceptibility. Most of these mutations are located in a highly diverse six amino acids window (*orange box*). In *D. radiodurans*, this window includes Arg144 (*green square*). **c** L3 protein penetrates deeply toward the vicinity of the PTC (*right image*, L3 in *green*, SB-571519 in *red*). The L3 nonconserve loop-like region (consisting of six amino acids, shown in **b** in the *orange box*) is colored *yellow* in the *center* and *left* image. In *D. radiodurans*, Arg144 electrostatically interacts with U2506 phosphate (**a**) 23S rRNA nucleotides 2504–2506 (*left, gray* surface representation) define a large portion of induced-fit binding pocket (see text), obtained mainly by conformational change of U2506. **d** Differences in remote nucleotides acquire pleuromutilin selectivity. An example is U2504, which is among the nucleotides that define pleuromutilins binding site. In eubacteria it points toward the PTC, whereas in the Archaeon H50S, this nucleotide stacks with 2055, which is A in eukarya and archaea but C in eubacteria. This stacking seems to stabilize U2504 position away from the binding pocket. SB-571519 is *red*; nucleotides 2504 and 2055 are *blue* in its complex with D50S and *purple* in H50S

were detected in protein L3 in all studied pleuromutilin derivatives as well as tiamulin [2].

3.2 Structural Basis for Cross Resistance Between Ribosomal PTC Antibiotics

The spatial distributions of nucleotides associated with resistance to PTC antibiotics identified by various non crystallographic methods (Table 2.2) mapped onto D50S crystal structure (Figs. 3.4 and 3.5), indicate that almost all of the nucleotides mediating resistance are clustered in a distinct region, although the PTC surroundings offer various nucleotides for antibiotics binding, some of which can be mutated without losing cell vitality [12]. This region is located farthest from the intersubunit interface and stretches into the entrance to the nascent protein exit tunnel (Figs. 3.4 and 3.5). The only outlier is the highly flexible nucleotide A2062, which is located closer to the subunit interface and was detected in different orientations in complexes of various antibiotics, namely chloramphenicol [13], streptogramin$_A$ [14], lankacidin [15] and methymycin [16].

For describing the locations of the nucleotides mediating antibiotics resistance, the PTC was divided by an artificial plane into two regions: one contains components participating in resistance mutations, and the other consists of nucleotides that are not involved in resistance. This artificial plane is defined by two perpendicular imaginary axes, namely the postulated 2-fold symmetry axis [6, 17] and the line connecting the bases of nucleotides G2553 and G2251 (Fig. 3.5), which are located at the boundaries of the space consumed by the rotatory motion of the translocating tRNA 3′ end from the A to the P site, and are engaged in Watson–Crick base pairing with both tRNAs (Fig. 3.5). Thus, the translocation of the tRNA required for nascent chain elongation is performed by two correlated motions: sideways progression of most of the tRNA molecule together with the mRNA by one codon at a time, and a rotatory motion of the aminoacylated 3′ end of the A site tRNA around the bond connecting it to the rest of the tRNA molecule. This bond coincides with the rotation axis of a pseudo-symmetrical region, comprising 180 nucleotides and located in and around the PTC [6, 17] within the otherwise asymmetric ribosome. When viewing from the subunit interface, the PTC wall associated with PTC antibiotics resistance is located behind the site of peptide bond formation (called the PTC "rear wall"). It is the main constituent of PTC region that navigates and guides the translocation of the tRNA 3′ end from the A to the P site, by creating extensive transient interactions of its backbone.

Figure 3.6a shows that for clinically relevant complexes, half of the nucleotides mediating resistance to PTC antibiotics are located at distances of 6–12 Å from the affected antibiotic. For instance, pleuromutilins resistance is acquired by mutations of remotely located nucleotides of the 23S RNA as well as of residues of r-protein L3 [2, 11, 18]. Figure 3.6b shows that approximately half of the nucleotides involved in resistance to PTC antibiotics affect two or more different families of

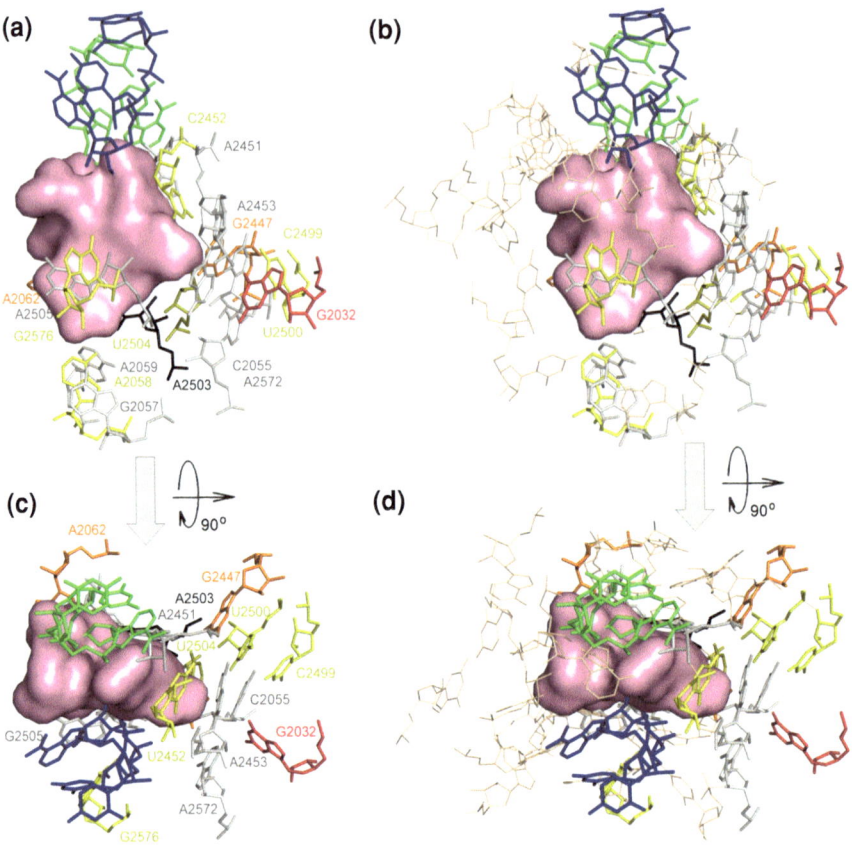

Fig. 3.4 Nucleotides shown to undergo mutations or methylations that confirm PTC antibiotic resistance or reduce susceptibility. Nucleotides are colored according to the number of affected classes (one, two, three, four, and five classes are represented by *gray*, *yellow*, *orange*, *red*, and *black*, respectively). The *pink* surface shows the total volume occupied by the PTC antibiotics: clindamycin (lincosamides), dalfopristin (streptograminsA), retapamulin (pleuromutilins), chloramphenicol (phenicols), and linezolid (oxazolidinones). For orientation, the A site tRNA 3′ end and the derived P site tRNA 3′ end are shown in *blue* and *green*, respectively. Images were taken from the direction of the L7/12 stalk (**a** and **b**) and from the top of the cavity leading to the PTC (**c** and **d**). Nucleotides located in the vicinity of the PTC within a distance < 8 Å from the corresponding antibiotic, and are not involved in known resistance determinants, are either shown as wheat-colored lines (**b** and **d**) or excluded (**a** and **c**)

antibiotics. In some cases resistance to drugs from up to five different antibiotic families are associated with same nucleotide (Table 2.2 and Fig. 3.6b).

Confining the resistance space to a single region in the PTC (Figs. 3.4 and 3.5) regardless of the bacterial species and the used methodology, indicates a functional distinction between the two sides of the PTC. Limiting the mutations to the PTC rear wall is consistent with the finding that in most cases the rear wall backbone, rather than its bases, plays a crucial functional role in guiding the

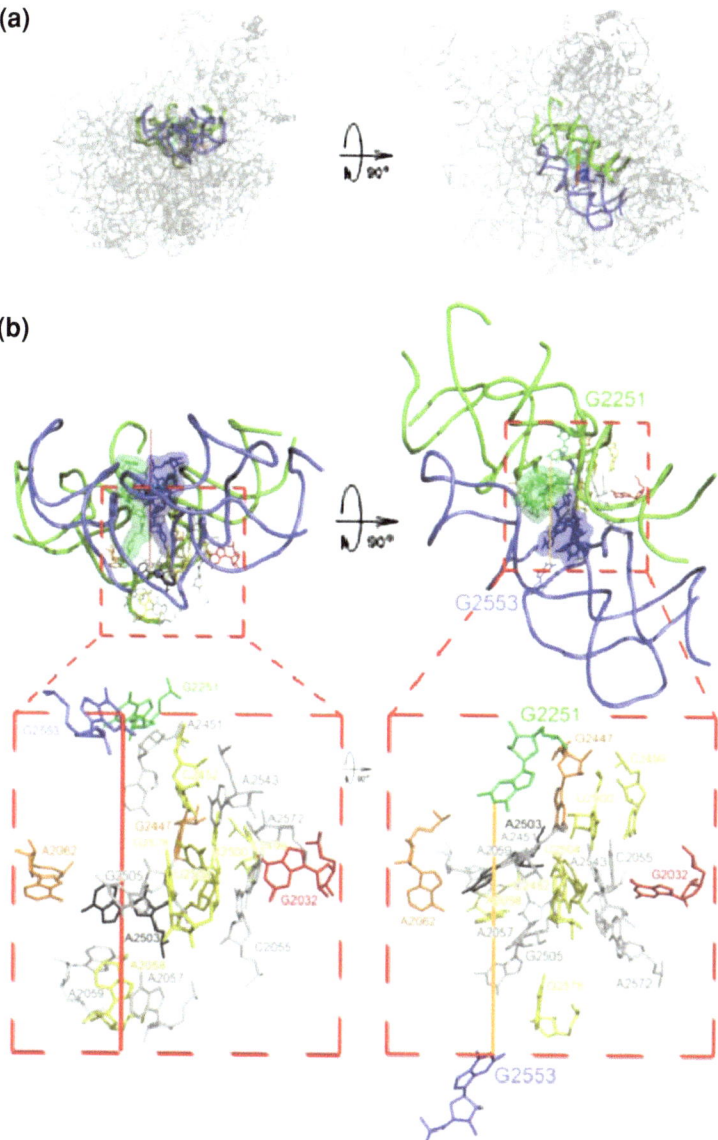

Fig. 3.5 The centrality of the symmetrical region and the 2 perpendicular axes that define the planes confining the PTC regions that participate in antibiotic resistance. In all: images were captured perpendicular to the 2-fold symmetry axes (*right*) and to the axes defined by the line connecting G2553 to G2251 (*left*). The rRNA backbone is shown in *gray*, except for the symmetrical region, where the subregion containing the A loop (called A region), is colored *blue* and that containing the P-loop (the P-region) is shown in *green*. The imaginary 2-fold rotation axis is shown in *red*. Nucleotides are colored according to the number of affected classes (one, two, three, four, and five classes are represented by *gray*, *yellow*, *orange*, *red*, and *black*, respectively). CCA 3′ ends of tRNA in the A and the P sites colored *blue* and *green*, respectively. **a** Two views of the large ribosomal subunit showing the centrality of the internal 2-fold

◀ **Fig. 3.5** (continued) symmetry region. The *left side* shows the large subunit as seen from its interface surface, and the *right side* is a view taken into the PTC. **b** The symmetrical region with the two imaginary axes relevant to cross-resistance. *Left* : The 2-fold symmetry axis colored *red*. *Right* : The line connecting G2553 to G2251, which divides the front and back walls of the PTC, colored *orange*. The 2-fold pseudo-symmetry regions of the PTC, including the A loop and the P loop, are *pale blue* and *green*, respectively. Zooms into these views, in which the nucleotides involved in resistance or reduced susceptibility are depicted, are shown in the *lower panels*

translocation of A site tRNA 3′ end from A to P site during peptide bond formation [6, 17].

3.3 Oligonucleotides as Ribosomal Inhibitors and as Tools for Structural and Functional Study

Twenty one target sites were selected from the 23S rRNA of the *E. coli* ribosome, based on structural, thermodynamical and genetic information, according to the following guidelines:

1. All target sites were limited to known functional centers, namely subunits interface, L7/12 and L1 stalks, the exit tunnel and the vicinity of the PTC.
2. A target site was preferred if it contained a mutation site that was previously related to antibiotic resistance, slow growth or lethality.
3. Tm (melting temperature) was set to be no less than 37°C in the presence of 1 µM ODN.
4. Intermolecular and intramolecular energy were set to be higher than −5.1 kcal/mol and −0.8 kcal/mol, respectively.
5. A target site was preferred if its total binding energy was relatively low.
6. Most exposed or most buried sites were selected.
7. ODNs length varied from 10 to 22 bases.

ODNs capability to inhibit protein translation of *E. coli* cell free system was assayed as described in methodology section above. The IC_{50} values were varied from 11 ± 2 to 148 ± 32 µM (Table 3.2).

3.3.1 Correlation between IC_{50} and Various ODN Parameters

Analysis of the magnitude of inhibition of all 21 oligonucleotides in cell free system showed statistically significant correlation between IC_{50} and ODN length ($r = -0.48685$, $p = 0.025206$), as well as with the potential number of h-bonds ($r = -0.50586$, $p = 0.019304$), i.e. the sum of $2*(\text{sum}(A) + \text{sum}(T)) + 3*(\text{sum}(G) + \text{sum}(C))$, where r is the Pearson correlation coefficient, p is the observed significance level of the test (P-value, namely the probability of

Table 3.2 IC_{50} of the various ODNs

ODN	IC_{50} [μM]	n
463-10	148 ± 32	3
570-17	33 ± 0	2
713-17	21 ± 6	4
817-20	41 ± 10	4
1042-14	24 ± 4	3
1067-22	47 ± 10	3
1064-15	46 ± 12	4
1064-12	56 ± 11	4
1082-15	75 ± 25	4
1082-12	18 ± 4	4
1655-17	57 ± 13	3
1681-15	28 ± 6	2
1909-18	46 ± n.a.	1
1923-12	116 ± 47	4
1989-17	11 ± 2	2
2306-15	113 ± 29	4
2444-14	54 ± 26	4
2515-14	94 ± 27	4
2653-15	40 ± 3	3
2654-13	50 ± 14	4
2655-11	94 ± 11	3

IC_{50} assayed based on the activity of *E. coli* (BL21(DE3) strain, Stratagene) cell-free transcription translation system in the presence of ODNs, with firefly luciferease as reporter. Standard deviation was derived based on the number of independent experiments as indicated on the right column

obtaining the observed correlation coefficient providing that the value of r is actually 0), sum(X) is the number of nucleotides of the type X in a given ODN. Other parameters (such as describing sequence conservation, calculated energy require for unwinding rRNA secondary structure, calculated energy gained by antisense binding, surface accessibility, number of hydrogen bonds between rRNA elements as observed crystallographically at the ribosomal target site, etc.) resulted in correlation coefficients that are not statistically significant.

3.3.2 Effect of ODN Length

Since the correlation observed between the extent of inhibition and ODN length, and with additional parameters linked to ODN length (e.g. 'number of potential h-bonds', namely the total number of donors and acceptors for Watson–Crick base pairing) the effect of ODN length was further examined. The oligonucleotides were split into two groups and correlation was recalculated for each group separately. One group included the relatively short sequences (length ≤ 14, n = 9) while the other group included the relatively long sequences (length > 14, n = 12). The cutoff at length of 14 bases was selected in order to obtain two groups with a similar number of ODNs (n).

3.4 Minimal Ribosomal Components with PTC Structure and Function

3.4.1 Construct Design

Constructs were designed in order to create a mimic of the minimal rRNA construct that resembles the surrounding of the PTC and is capable of performing of peptide bond formation. Constructs were designed based on nucleotides sequence within parts or all of the symmetrical region within the 23S rRNA of *T. thermophilus*, which is identical to *D. radiodurans* (see Fig. 3.7 and Table 3.3).

Minimal constructs contained two helices, corresponding to the 23S rRNA H74 and H89 (named P1) or H90 and H93 (named A1) and the elbows connecting

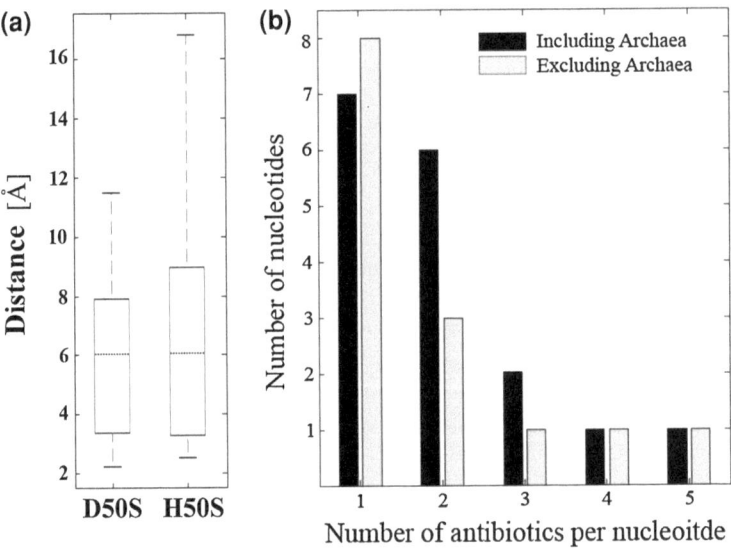

Fig. 3.6 Overlaps of resistance determinants and distances between antibioticsbinding sites and nucleotides mediating resistance. **a** Boxplot representation ofnucleotide-antibiotic distances determined for D50S and H50S complexes. *Dotted lines*show medians; the *upper* and *lower horizontal lines* of the boxes stand for upper andlower quartiles (namely cutoffs for 25% and 75% of the data). *Top* and *bottom* external*horizontal lines* show the maximal and minimal values. **b** Resistant mutations ormethylations observed in bacterial strains including (*black bars*) or excluding (*gray bars*)archaea. x axis indicates the number of different classes of antibiotics that are beingaffected. y axis stands for the number of nucleotides characterized for this observation.

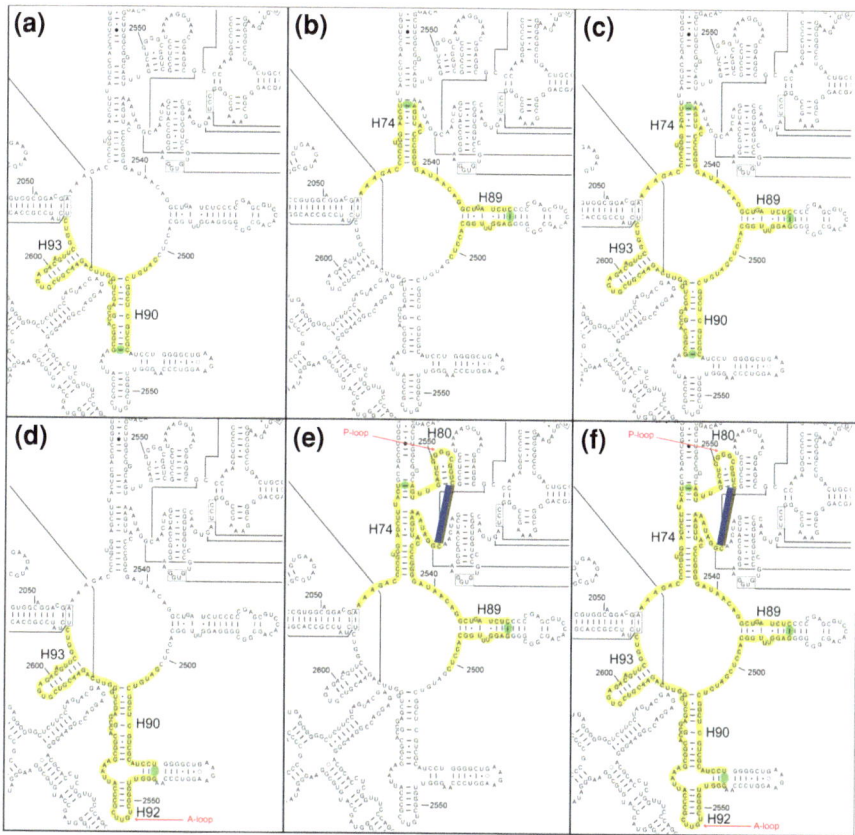

Fig. 3.7 Two-dimensional representation of 23 rRNA of *T. thermophilus* from the surroundings of the PTC. Nucleotides highlighted *yellow* in **a–f** were used to design the sequences of RNA constructs A1, P1, AP1, A2, P2 and AP2, respectively. *Blue stretches* indicate direct connection between nucleotides and *green dots* represents positions where the hyper stable CUUCGG hairpin was inserted. This was done in order to avoid rRNA sequences that perturbing from the A- or P-regions and do not obey the 2-fold symmetry within the surroundings of the PTC, according to the three-dimensional structure. rRNA helices and nucleotide numbers (*E. coli* numbering) are designated in *black*. A-loop and P-loop are pointed by *red arrows*

them. These two pairs of helices form the major part of the 2-fold symmetrical region, in which H74 and H89 are symmetrically related to H90 and H93 [17, 19]. The first nucleotide in P1 corresponds to nucleotide 2058. This nucleotides is a conserved A in eubacteria. In the construct it was replaced to G in order to meet the requirement for a leading guanosine in T7 RNAP transcript [20]. This alteration is not expected to result in a negative effect since this nucleotide is a conserved G in eukaryotes and archaea. Short sequences of up to six nucleotides were inserted in order to form a stable hairpins within positions where original rRNA helixes, which are longer in the contemporary ribosome, were truncated to form

Table 3.3 RNA sequences. Construct names appear as referred within the text and figures

Name	Sequence
A1	GAUGUCGGCUCGCUCGCUCGGCGGCCACGCGAGCUGGGUUCAGAACGUCGUGAGACAGUUCGGUC
A2	GAUGUCGGCUCGCUCGCUCGGCAUCCUUCGGGUUGGGCGUGUUCGCCCAUUAAAGCGGCUGGGUUCAGAACGUCGUGA GACAGUUCGGUC
AP1	GAAGACCCGUGGAGCUCUUCGGAGUUACCCCGGGAUAAACAGGCUGAUCUCUUCGGAGGUUUGGCACCUCGAUGUCGGC UCGUCGCUCGGCGGCCACGCGAGCUGGGUUCAGAACGUCGUGAGACAGUUCGGUC
AP2	GAAGACCCGUGGAGCUUUACUCUUCGGAGUUGACUGGGCGGUCCGAUAAAAGUUACCCCGGGAUAACAGGCUGAU CUCUUCGGAGGUUUGGCACCUCGAUGUCGGCUCGUCGCUCGAUUCGGCUCAUCCUU CGGGUUGGGCGUUCGCCCAUUAAAGCGGCACGCGAGCUGGGUUCAGAACGUCGUGAGACAGUUCGGUC
P1	GAAGACCCGUGGAGCUCUUCGGAGUUACCCCGGGAUAACAGGCUGAUCUCUUCGGAGGUUUGGCACCUC
P1.del	GAAGACCCGUGGAGCUCUUCGGAGUUACCC CGGGAUAACAGGCUGAUCUCUUCGGGGUUUGGCACCUC
P1.ins	GAAGACCCGUGGAGCUCUUCGGAGUUACCCCGGGAUAACAGGCUGAUCUCGGCCAGGUUGGCACCUC
P1c	GAAGACCCGUGGAGCUCUUCGGAGUUACC CCGGGAUAACAGGCUGAUCGUGAGGUUUGGCACCUC
P1c.G53U	GAAGACCCGUGGAGCUCUUCGGAGUUACCCCGGGAUAACAGGCUGAUCGUUAGGUUUGG CACCUC
P1d	GAAGACCCGUGGAGCUCUUCGGAGUUACCCCGGGAUAACAGGCUGGGUCGUGAGACGGCACCUC
P2	GAAGACCCGUGGAGCUUUACUCUUCGGAGUUGACUGGGCGGUCCGAUAAAAGUUACCC CGGGGAUAUAACAGGCUGAU CUCUUCGGAGGUUUGGCACCUC
P2c	GAAGACCCGUGGAGCUUUACUCUUCGGAGUUUGACUGGGCGGUCCGGAUAAAAAGUUACCCCGGGAUAUAACAGGCUGAU CGUGAGGUUUGGCACCUC
P2d	GAAGACCCGUGGAGCUUUACUCUUCGGAGUUUGACUGGGCGGUCCGGAUAAAAAGUUACCCCGGGAUAUAACAGGCUGGU CGUGAGACGGCACCUC

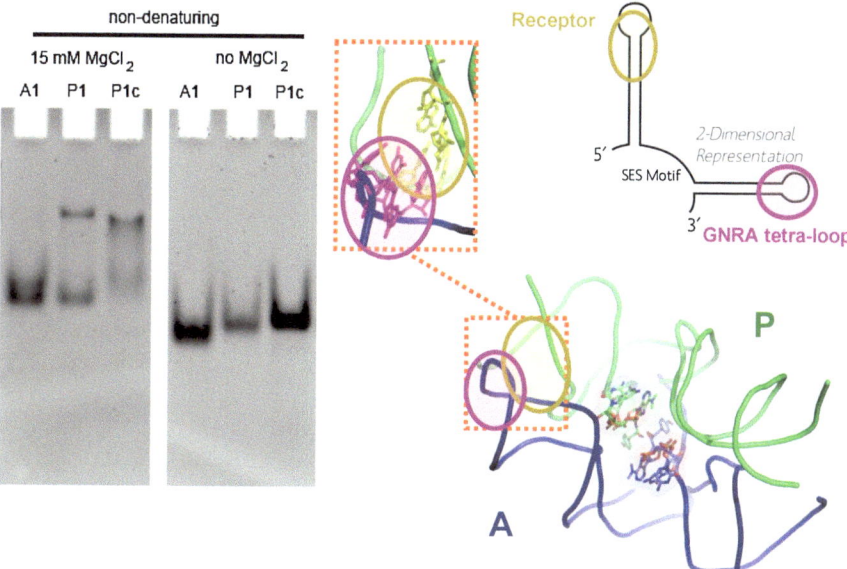

Fig. 3.8 Incorporation of GNRA tetra-loop to P1 RNA sequence resulted with enhanced dimerization. *Bottom right*: Three-dimensional backbone representation of the minimal component of the 2-fold symmetry region from the surroundings of the PTC, within the crystal structure of the large ribosomal subunit from the eubacteria *D. radiodurans*. A-region and P-region marked accordingly and colored *blue* and *green*, respectively. Coordinates of CCA 3' ends in the PTC are surrounded with transparent surface representation and colored as *CPK*, with carbon atoms in *blue* and *green* for substrates located in the A- and the P-site, respectively. *Dashed orange box* highlights the location of A-minor interactions between the GNRA (GUGA) tetra-loop located in H93 (*purple circle*) and the receptor at H74 (*yellow circle*). *Top middle dashed box*: Zoom into the residues involved in A-minor interactions, in sticks representation. The four nucleotides in the GUGA tetraloop are colored *purple* and nucleotides forming the receptor are colored *yellow*. *Top right*: Two-dimensional representation of P1c construct. The construct was designed to form two helices baring the sequence of nucleotides of the P-region, as shown at the *bottom right* panel. *Yellow circle* indicates location of the receptor, designed based on the corresponding H74 within the contemporary ribosome. *Purple circle* indicates the location of the GNRA tetra-loop that was taken from H93 within the sequence of contemporary 23S rRNA and inserted into P1c within the helix corresponding to H89. *Left*: Non-denaturing PAGE of A1, P1 and P1c (lanes marked accordingly) in the presence (*left gel*) and absence (*right gel*) of magnesium ions (see methodology and results sections for full description)

minimal constructs that include only the symmetrical region [17]. These inserted loops were designed based on the stable hexamer hairpin 5'-CUUCGG-3' [21, 22], a pattern found in various mRNA sequences in the center of palindrome-like patterns that were predicted to form stem loops [22]. This pattern is also found in the 16S rRNA (starting in positions 207, 419, 1028, 1449) [22] and the 23S rRNA (starting at nucleotide 1691), where it folds as loops at the tips of a well defined rRNA helices in all ribosome crystal structures.

In the contemporary ribosome H80 and H97 carry the P-loop and the A-loop, respectively. These highly conserved loops reside within the symmetrical region,

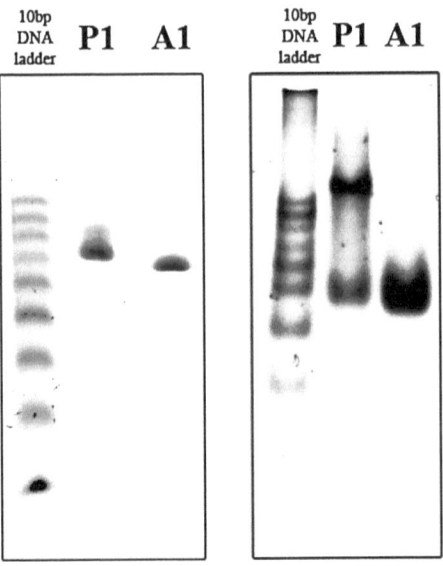

Fig. 3.9 Denaturing and non-denaturing polyacrylamide gel electrophoresis of RNA constructs A1 and P1. *Left*: 15% polyacrylamide gel under denaturing conditions. Both A1 and P1 migrate according to their molecular weight. The 10 bp DNA ladder composes of 10–100 bp DNA segments. *Right*: 12% polyacrylamide under non-denaturing conditions. Each loaded sample contained RNA in 2 μM concentration (see methodology for full description). While A1 migrates with only one band, two bands were clearly observed for P1, indicating a fast migrating monomer and a slow migrating dimer

and form Watson–Crick base-pairs with the 3′ CCA end of tRNA substrates within the contemporary ribosome. Sequences of these loops, together with rRNA sequences of the corresponding helices that carry them, were incorporated into the sequence of P1 and A1 to form larger constructs, namely P2 and A2 respectively, each bearing the entire nucleotides sequence of either the full A- or P-regions within the contemporary ribosome (Fig. 3.7, Table 3.3). Additional constructs were designed based on the sequence of the entire 2-fold symmetrical region in the absence or presence of H80 and H90 and named AP1 and AP2, respectively (Fig. 3.7, Table 3.3).

In the contemporary ribosome the interactions between the two parts of the symmetrical region involve an A-minor interaction [23] between GNRA tetraloop, where N is any nucleotide and R is purine, in H93 at the A-region (including the GNRA tetraloop sequence GUGA) and a receptor located at H74 in the P-region (Fig. 3.8). The sequence of H74 already included in both RNA constructs P1 and P2. Therefore, I have incorporated the GUGA sequence into these constructs, at the location corresponding to H89, instead of the hyper stable hairpin CUUCGG (constructs named P1c and P2c). This design was applied in order to enable the formation of P1 or P2 homodimers with enhanced affinity between monomers and probability for improved organization of a tightly packed dimer forming a pocket,

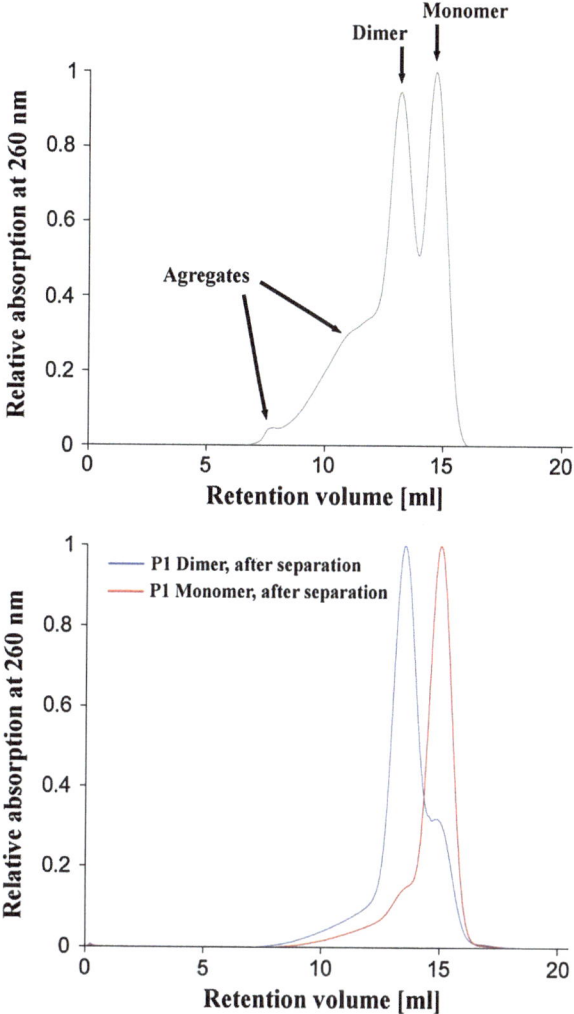

Fig. 3.10 Separation of P1 dimer and monomer by size exclusion chromatography. *Top*: Chromatogram obtained following injection of 100 μg P1 RNA in 300 μM concentration after incubation in conditions that permit dimerization. Fractions including the dimer and monomer (indicated) were collected and injected for a second run (see text for full details). *Bottom*: Chromatograms obtained in the second run, after injection of the monomer (*red*) and the dimer (*blue*) fractions

rather than extended arrangement, based on symmetrical considerations. As it was impossible to predict if the inclusion of the GUGA tetra-loop per se is sufficient and optimal to promote correct fold and dimerization, another two constructs were derived from P1c and P2c to carry the GUGA together with the addition of three adjacent base pairs from H93, named P1d and P2d (Table 3.3).

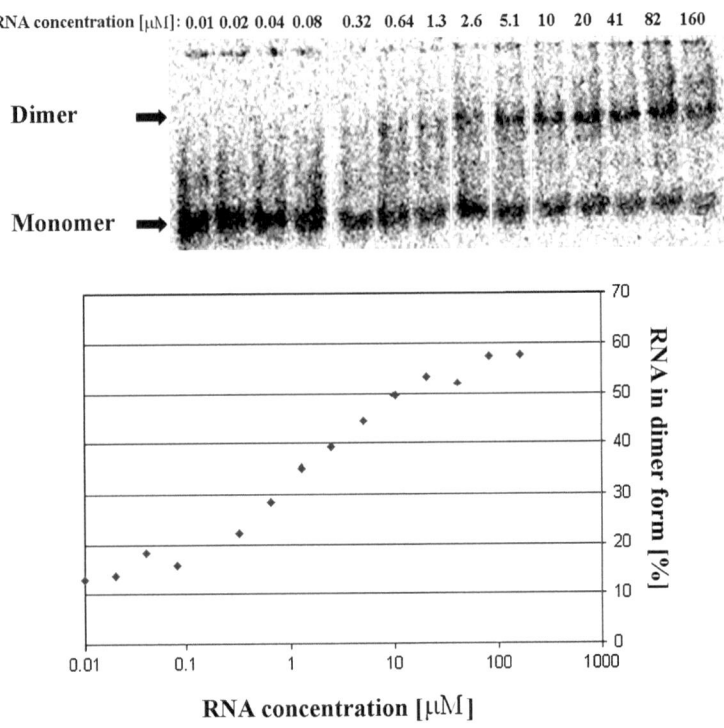

RNA concentration [μM]: 0.01 0.02 0.04 0.08   0.32 0.64 1.3   2.6   5.1   10   20   41   82   160

Dimer ➡

Monomer ➡

Fig. 3.11 RNA concentration dependent P1 homodimer formation. *Top*: EMSA performed using ^{33}P labeled P1 RNA (see text for further details). *Bottom*: The amount of the low migrating band is concentration dependent, indicating for a homo-dimer with K_d of the order of 10^{-5} M

3.4.2 Study of Dimerization Tendency

In order to test the hypothesis that the 2-fold symmetrical region surrounding the PTC evolved by gene fusion or gene duplication, I tested its tendency to form dimers, the prerequisite for obtaining the conformation required for supporting peptide bond formation by the two components, namely the A1 and P1 RNA constructs. Accordingly, mixtures of the two constructs, A1 and P1, were subjected to gel electrophoresis under denaturing and non-denaturing conditions. A single band was detected for A1 in both denaturing and non-denaturing gels, indicating for no dimerization under the experimental conditions (Fig. 3.9). However, For P1, one band was detected in denaturing gel but two bands appeared in the non-denaturing gel, hence indicating for P1 dimerization.

I have further verified this observation by Size Exclusion Chromatography (SEC) Chromatogram obtained following injection of 100 μg P1 RNA in 300 μM concentration indicated for coexistence of monomer, dimer and some higher order aggregates (Fig. 3.10, top panel). Approximately 750 μl fractions of both dimer

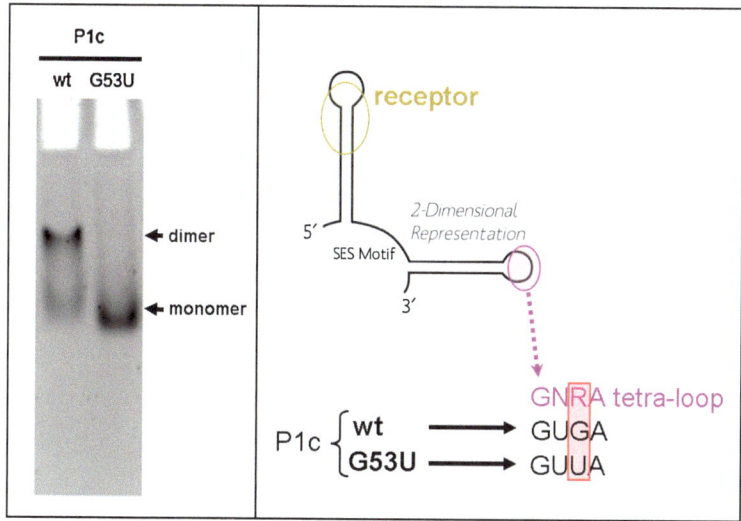

Fig. 3.12 Point mutation in P1c to abolish the GNRA motif resulted with loss of dimerization. *Right*: Two-dimensional representation of P1c design includes the location of the receptor (*yellow circle*) and the GNRA tetra-loop (*purple circle*). The mutation G53U is highlighted with a *red box*. *Left*: Non-denaturing PAGE of P1c wild type and G53U mutant. Bands of dimer and monomer are indicated

and monomer were collected in 1 μM concentration. Fractions were kept on ice for > 30 min, and then were injected again to the same column without any further treatment in order to evaluate the quality of the separation and to assess the thermodynamic stability of the dimer and monomer for the time required for optional further enzymatic characterization [24]. Under such experimental conditions the enzymatic reaction can be immediately performed, before equilibrium between dimer and monomer is achieved, and hence detect even metastable RNA structures [25]. Both chromatograms indicated a significant excess of the desired mer over the alternative entity, with a remarkably reduced amount of aggregates (Fig. 3.10, bottom panel), indicating that both dimer and monomer were thermodynamically stable under the experimental conditions.

The detection of two bands on non-denaturing gel can be an indication for two RNA conformers, rather than for a dimer and monomer. Therefore, using EMSA P1 was incubated at elevating RNA concentrations (10 nM to 160 μM), and then subjected to non-denaturing gel electrophoresis. The concentration-dependence of the slow migrating band that was observed indicates a homo-dimer formation with a Kd of $\sim 10^{-5}$ M (Fig. 3.11). This result supports the hypothesis that the 2-fold symmetry region could have been potentially formed by the dimerization of one of its asymmetrical components.

Based on these results I focused on constructs designed according to P1 sequence. In an attempt to increase the tendency to dimerize, I incorporated a GNRA tetra-loop into the helix corresponding to H89 to form the RNA construct

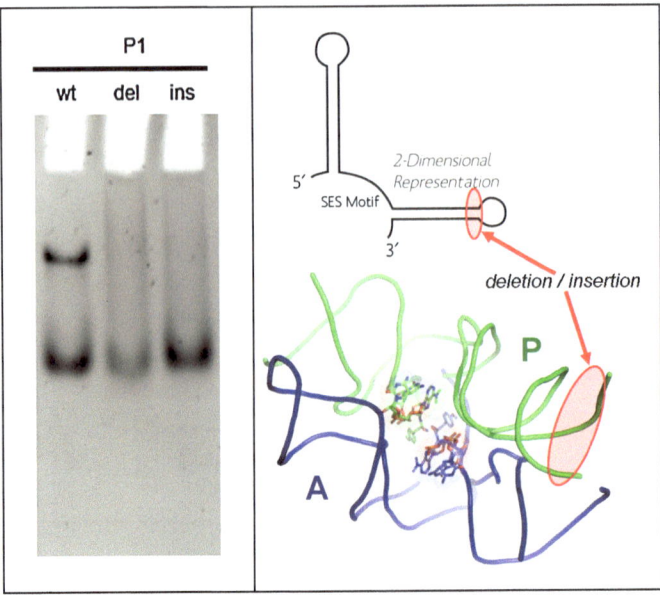

Fig. 3.13 Dimerization of P1 RNA abolished following either insertion or deletion of two nucleotides designed to form a single base-pair. *Right*: Three-dimensional representation of rRNA at the surroundings of the PTC includes the minimal components within the A-region and the P-region required for the formation of the interface between the two symmetrical regions and the PTC catalytic center (*bottom*) and two-dimensional representation of these constructs, as designed (*top*). Sequences of A1 and P1 RNA were designed based on the sequence of 23S rRNA nucleotides represented here in *blue* and *green*, respectively. Coordinates of CCA 3′ ends in the PTC are surrounded with transparent surface representation and colored as *CPK*, with carbon atoms in *blue* and *green* for substrates located in the A- and the P-site, respectively. Ellipsoids represent for the location where a single base-pair was either inserted or deleted. *Left*: Non-denaturing PAGE for P1 (designated "wt") and the two double mutants including either base-pair deletion or insertion (designated "del" and "ins", respectively)

P1c (Fig. 3.8, right). P1c formed a dimer with higher affinity then P1, as can be seen by the increased intensity of the slaw migrating band that represents the dimer, in respect to the fast migrating band of the monomer (on non-denaturing PAGE in the presence of magnesium) (Fig. 3.8, left). In order to verify the contribution of the GNRA tetra-loop to dimer formation I introduced to P1c a "point mutation" G53U, thus abolishing the GNRA motif by converting the GUGA tetra-loop sequence to GUUA (Fig. 3.12, right). As a result, the "mutant" failed to form a dimer (Fig. 3.12, left), indicating that the GNRA sequence contributes to dimer formation, presumably through A-minor interactions.

Notably, the "mutation" G53U in P1c did not shifted the thermodynamic equilibrium between dimer and monomer to the reduced level observed in case of P1, but completely abolished dimerization to below detection limit. The minor sequence alteration between P1c and P1 at the helix corresponding to H89 in the contemporary ribosome (see Fig. 3.8 and text above) is limited to nucleotides

Fig. 3.14 EMSA for the detection of heterodimerization between A1, P1 and P1c. Experiments performed as described in methodology above. Non-labeled RNA concentration in all samples was 1 µM. Non-labeled RNA construct used in each sample is indicated above each lane and ^{32}P radio-labeled used for each experiment is indicated at the left of the corresponding gel scan. "D" and "M" are for dimer and monomer, respectively

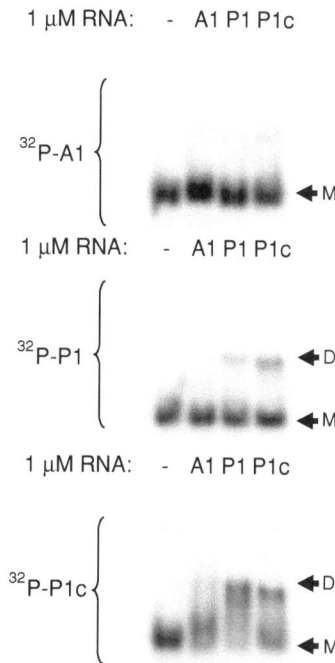

designed to be located on the opposite side of this helix with respect to the presumed catalytic site. Therefore, a plausible scenario upon GNRA motif abolishment would predict that affinity between both monomers will be similar to the affinity observed between monomers in case of P1, rather than completely eliminated. As this is not the case the complete loss of dimerization of P1c "mutant" G53U, may indicate a more general role of the remote end of this helix in promoting dimerization also in case of P1. The contribution of the corresponding region of P1 to dimer formation is well supported by the loss of dimerization observed when either insertion or deletion of a designed single base-pair were introduced to the corresponding location of P1 (namely the remote end of H89 in the contemporary ribosome) (Fig. 3.13).

I tested the ability of RNA constructs A1, P1 and P1c to form heterodimers by EMSA (Fig. 3.14). While A1 failed to form heterodimers with either P1 or P1c, the latter two confirmed to form a stable heterodimer presumably via A-minor interactions between the receptor in P1 to the GNRA tetra-loop in P1c. An explanation for failure of A1 to form either hetero or homo dimers can be the misfolding of the RNA construct, although it was designed based on rRNA sequence that is naturally folded and stabilized within the contemporary ribosome. This hypothesis can be supported by inaccurate two-dimensional diagram that was predicted for A1 by mFold [26] (Fig. 3.15), with compared to the structure observed for the corresponding region within the contemporary ribosome. In contrast to A1 folding prediction, both P1 and P1c were predicted to fold into a

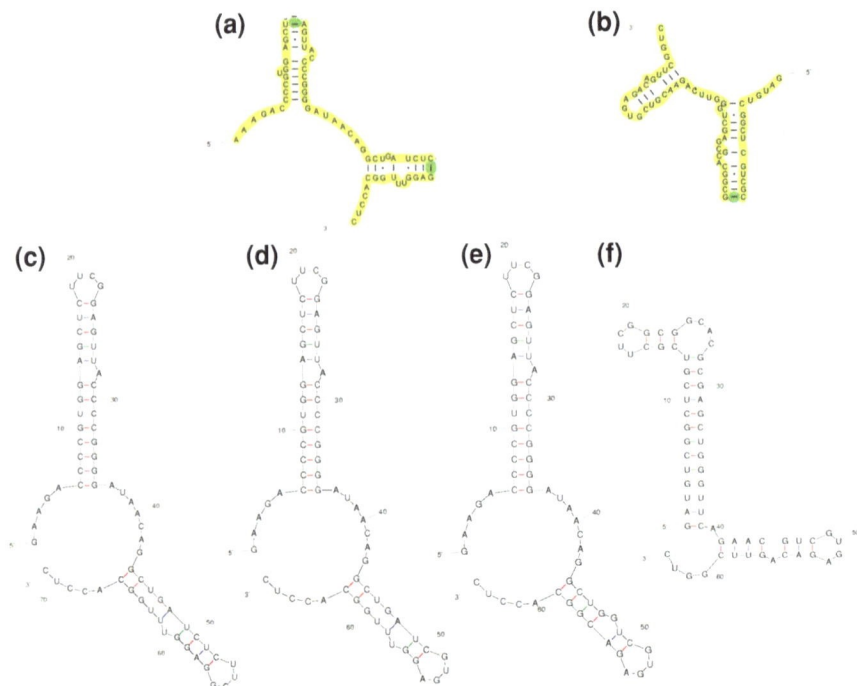

Fig. 3.15 Two-dimensional structure of RNA constructs P1, P1c and A1 as predicted by mFold. *Top*: Two-dimensional representation of nucleotides sequence within 23S rRNA that were used to design RNA constructs P1, P1c, P1d (**a** for all) and A1 (**b**) *Bottom*: Two-dimensional structures for P1, P1c, P1d and A1, as predicted by mFold, are in **c–f**, respectively

structure resembling their corresponding rRNA structure used for its design (Fig. 3.15).

Within the contemporary ribosome, the A- and P-loops are part of the symmetrical region that surrounds the PTC, but these loops are not involved in the actual positioning of very short substrates, such as puromycin derivatives (used for the fragment reaction), which may resemble the initial substrates. Based on this observation we have hypothesized that incorporation of the P-loop into either P1 or P1c constructs will not prevent their dimerization, assuming that these constructs indeed fold and dimerize into a structure resembling rRNA at the surroundings of the contemporary PTC. Accordingly, I incorporated nucleotides sequence of the P-loop, including H80 that carries it (Figs. 3.7 and 3.16), into the sequence of A1, P1 and P1c, resulting with RNA constructs A2, P2 and P2c, respectively (Table 3.3). Non-denaturing PAGE confirmed that both P2 and P2c formed homodimers (Fig. 3.16), as could be expected in case of a properly folded and assembled dimers. Another RNA constructs that were tested are P1d and P2d. These were derived from P1c and P2c by incorporation of ten nucleotides sequence from H93 in the contemporary ribosome to P1 and P1c RNA sequence within the region designed to form a helix corresponding to H89 in the contemporary ribosome. This

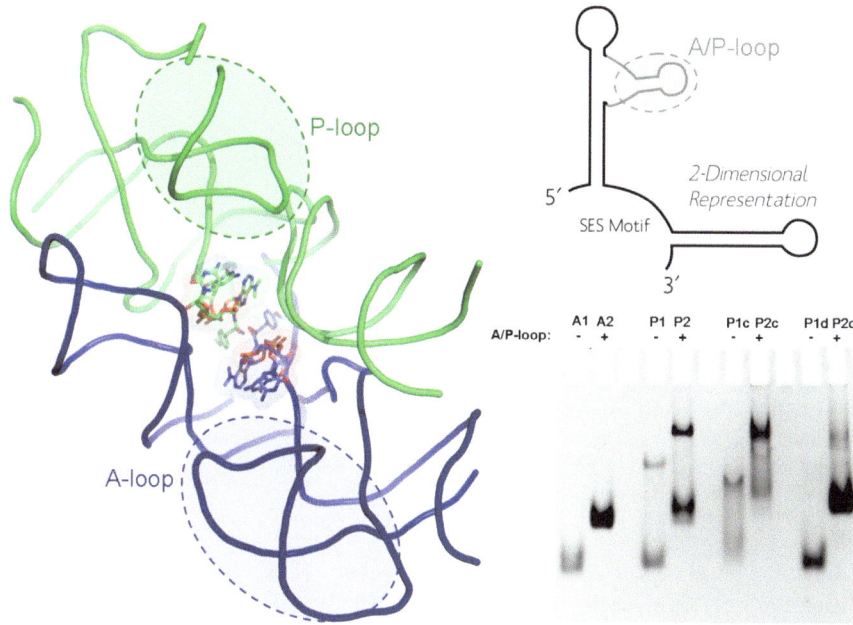

Fig. 3.16 Incorporation of the P-loop into P1 and P1c RNA did not prevent dimerization. *Left*: Three-dimensional structure of nucleotides from the surroundings of the PTC, as determined within the crystal structure of the large ribosomal subunit of the eubacteria *D. radiodurans*. A-region and P-region are colored *blue* and *green*, respectively. Coordinates of CCA 3' ends in the PTC are surrounded with transparent surface representation and colored as *CPK*, with carbon atoms in *blue* and *green* for substrates located in the A- and the P-site, respectively. *Blue* and *green dashed ellipsoids* highlight the location of the A- and P-loops, including the helices that carry them respectively. *Top right*: Two-dimensional schematic representation of the RNA constructs A2, P2 P2c and P2d as designed with the location of the A- or P-loops, including the helix carries it, highlighted by *dashed gray ellipsoid*. *Bottom right*: Non-denaturing PAGE of various RNA constructs, as indicated above each lane, that including ("+") or excluding ("−") the sequences of A- and P-loops and sequences of helices that carry them within the contemporary ribosome

sequence included the GUGA tetraloop and three nucleotides up and down stream of it. No significant evidence for remarkable dimerization observed under the experimental condition tested for these two constructs, apart for a very pale band for dimer observed in case of P2d (Fig. 3.16).

3.4.3 Functional Characterization: Assay for Peptidyl Transferase Activity

In order to test the functional capabilities of RNA constructs that formed dimers I performed a serial of peptidyl transferase assays using minimal substrates (see Fig. 3.17 for typical experiment results). Based on results obtained from the

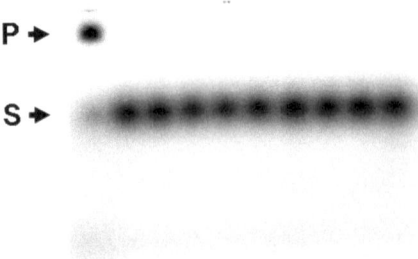

Fig. 3.17 Gel scan revealing results of a typical peptidyl transferase assay. Fast migrating band carries the unconsumed radiolabeled substrate (marked "S") ^{33}P-CCPmn. Slow migrating band including the streptavidin retarded product (marked "P"), ^{33}P-CCPmn-phenyl alanine-caproic acid-biotin. Product was generated by large ribosomal subunits from *E. coli* that served as a control throughout all experiments

dimerization assays, I tested the following pairs of RNA constructs (final concentration of 1 μM each): P1:P1, P1c:P1c, P1:P1c, P2:P2, P2:P2c, P2c:P2c and P2c:P2d. In addition, I tested 1 μM of the constructs AP1 and AP2. Each experiment included a control containing 1 μM of *E. coli* 50S ribosomal subunit instead of RNA constructs.

I tested various incubation conditions, with pH ranging from 6.0 to 8.0, at room temperature, 37 or 50°C, in the presence or absence of 30% methanol, using incubation periods of up to 24 h. In none of these experiments a detectable catalytic activity was observed. In all experiments the *E. coli* 50S subunits control showed catalytic activity, indicating that product detection was experimentally valid. While it is impossible to exclude the possibility that lack of catalytic activity could potentially result from insufficient variation of experimental conditions, yet another plausible explanation could be deficient or inaccurate fold, dimer formation and/or substrates binding and positioning. Any of these potential problems could have emerged because these RNA constructs carry rRNA sequence from the contemporary ribosome. It is possible that these sequences were evolved to fulfill their present task within the context of the ribosome. Therefore, when presented as an isolated entity it is possible that slight or extensive sequence alterations are required for enabling activity. Also, it is likely that the substrates used so far do not mimic the original substrates. Therefore, experiments with additional substrates will follow within the group.

References

1. Yan K, Madden L, Choudhry AE, Voigt CS, Copeland RA, Gontarek RR (2006) Biochemical characterization of the interactions of the novel pleuromutilin derivative retapamulin with bacterial ribosomes. Antimicrob Agents Chemother 50(11):3875–3881

2. Schluenzen F, Pyetan E, Fucini P, Yonath A, Harms J (2004) Inhibition of peptide bond formation by pleuromutilins: the structure of the 50S ribosomal subunit from *Deinococcus radiodurans* in complex with tiamulin. Mol Microbiol 54(5):1287–1294

3. Gindulyte A, Bashan A, Agmon I, Massa L, Yonath A, Karle J (2006) The transition state for formation of the peptide bond in the ribosome. Proc Natl Acad Sci U S A 103(36):13327–13332

4. Egger H, Reinshagen H (1976) New pleuromutilin derivatives with enhanced antimicrobial activity. II. Structure–activity correlations. J Antibiot (Tokyo) 29(9):923–927

5. Agmon I, Amit M, Auerbach T, Bashan A, Baram D, Bartels H, Berisio R, Greenberg I, Harms J, Hansen HA, Kessler M, Pyetan E, Schluenzen F, Sittner A, Yonath A, Zarivach R (2004) Ribosomal crystallography: a flexible nucleotide anchoring tRNA translocation, facilitates peptide-bond formation, chirality discrimination and antibiotics synergism. FEBS Lett 567(1):20–26

6. Bashan A, Zarivach R, Schluenzen F, Agmon I, Harms J, Auerbach T, Baram D, Berisio R, Bartels H, Hansen HA, Fucini P, Wilson D, Peretz M, Kessler M, Yonath A (2003) Ribosomal crystallography: peptide bond formation and its inhibition. Biopolymers 70(1):19–41

7. Harms J, Schluenzen F, Zarivach R, Bashan A, Gat S, Agmon I, Bartels H, Franceschi F, Yonath A (2001) High resolution structure of the large ribosomal subunit from a mesophilic eubacterium. Cell 107(5):679–688

8. Schuwirth BS, Borovinskaya MA, Hau CW, Zhang W, Vila-Sanjurjo A, Holton JM, Cate JHD (2005) Structures of the bacterial ribosome at 3.5 Å resolution. Science 310(5749):827–834

9. Selmer M, Dunham CM, Murphy FVt, Weixlbaumer A, Petry S, Kelley AC, Weir JR, Ramakrishnan V (2006) Structure of the 70S ribosome complexed with mRNA and tRNA. Science 313(5795):1935–1942

10. Bosling J, Poulsen SM, Vester B, Long KS (2003) Resistance to the peptidyl transferase inhibitor tiamulin caused by mutation of ribosomal protein l3. Antimicrob Agents Chemother 47(9):2892–2896

11. Pringle M, Poehlsgaard J, Vester B, Long KS (2004) Mutations in ribosomal protein L3 and 23S ribosomal RNA at the peptidyl transferase centre are associated with reduced susceptibility to tiamulin in Brachyspira spp. isolates. Mol Microbiol 54(5):1295–1306

12. Sato NS, Hirabayashi N, Agmon I, Yonath A, Suzuki T (2006) Comprehensive genetic selection revealed essential bases in the peptidyl-transferase center. Proc Natl Acad Sci U S A 103(42):15386–15391

13. Schluenzen F, Zarivach R, Harms J, Bashan A, Tocilj A, Albrecht R, Yonath A, Franceschi F (2001) Structural basis for the interaction of antibiotics with the peptidyl transferase centre in eubacteria. Nature 413(6858):814–821

14. Harms J, Schluenzen F, Fucini P, Bartels H, Yonath A (2004) Alterations at the peptidyl transferase centre of the ribosome induced by the synergistic action of the streptogramins dalfopristin and quinupristin. BMC Biol 2(1):4;1–10

15. Auerbach T, Mermershtain I, Davidovich C, Bashan A, Belousoff M, Wekselman I, Zimmerman E, Xiong L, Klepacki D, Arakawa K, Kinashi H, Mankin AS, Yonath A (2010) The structure of ribosome–lankacidin complex reveals ribosomal sites for synergistic antibiotics. Proc Natl Acad Sci U S A 107(5):1983–1988

16. Auerbach T, Mermershtain I, Bashan A, Davidovich C, Rosenberg H, Sherman DH, Yonath A (2009) Structural basis for the antibacterial activity of the 12-membered-ring mono-sugar macrolide methymycin. Biotechnolog 84:24–35

17. Agmon I, Bashan A, Zarivach R, Yonath A (2005) Symmetry at the active site of the ribosome: structural and functional implications. Biol Chem 386(9):833–844

18. Ippolito JA, Kanyo ZF, Wang D, Franceschi FJ, Moore PB, Steitz TA, Duffy EM (2008) Crystal structure of the oxazolidinone antibiotic linezolid bound to the 50S ribosomal subunit. J Med Chem 51(12):3353–3356

19. Agmon I, Auerbach T, Baram D, Bartels H, Bashan A, Berisio R, Fucini P, Hansen HA, Harms J, Kessler M, Peretz M, Schluenzen F, Yonath A, Zarivach R (2003) On peptide bond formation, translocation, nascent protein progression and the regulatory properties of ribosomes. Eur J Biochem 270(12):2543–2556

20. Christopher WS (1998) RNA:protein interactions: a practical approach, 1st edn. Oxford University Press, Oxford

21. Cheong C, Varani G, Tinoco I Jr (1990) Solution structure of an unusually stable RNA hairpin, 5'GGAC(UUCG)GUCC. Nature 346(6285):680–682

22. Tuerk C, Gauss P, Thermes C, Groebe DR, Gayle M, Guild N, Stormo G, d'Aubenton-Carafa Y, Uhlenbeck OC, Tinoco I Jr et al (1988) CUUCGG hairpins: extraordinarily stable RNA secondary structures associated with various biochemical processes. Proc Natl Acad Sci U S A 85(5):1364–1368

23. Nissen P, Ippolito JA, Ban N, Moore PB, Steitz TA (2001) RNA tertiary interactions in the large ribosomal subunit: the A-minor motif. Proc Natl Acad Sci U S A 98(9):4899–4903

24. Knapp G (1989) Enzymatic approaches to probing of RNA secondary and tertiary structure. Methods Enzymol 180:192–212

25. Linnstaedt SD, Kasprzak WK, Shapiro BA, Casey JL (2006) The role of a metastable RNA secondary structure in hepatitis delta virus genotype III RNA editing. RNA 12(8):1521–1533

26. Zuker M (1989) On finding all suboptimal foldings of an RNA molecule. Science 244(4900): 48–52

Chapter 4
Discussion

4.1 Pleuromutilins

4.1.1 Induced-Fit Mechanism for Pleuromutilin Binding

U2506 and U2585 undergo the most significant rRNA structural rearrangements upon binding of all three pleuromutilins. The conformational alterations of U2506 close tightly the binding pocket on the bound tricyclic core of all pleuromutilins (Fig. 3.1g–i). The altered conformation of U2506 is stabilized by its hydrophobic interactions with the bound compound as well as by its interactions with U2585, which shifts away from the C14 extension. These interactions may account for the protection of N3 in U2506 and U2585 detected by chemical footprinting experiment done by 1-cyclohexyl-3-(2-morpholinoethyl) carbodiimide metho-*p*-toluenesulfonate (CMCT) in *E. coli* and *B. hyodysenteriae* in the presence of tiamulin, valnemulin, pleuromutilin, and the carbamate derivative SB-264128 (Fig. 1.2) [1–4]. It is worth noting that U2506 and U2585 motions are not conditional for pleuromutilins binding because slightly different location of the compound can partially compensate for it, as observed for tiamulin, an additional C14-sulfanyl acetate derivative [5].

U2506 and U2585 nucleotides were identified as essential for ribosomal function biochemically [6] and by systematic genetic selection [7]. Both nucleotides are highly flexible, as noticed due to the differences of their orientations in crystals kept in environments that are close [8] or far [9] from physiological conditions. Consistently, conformational rearrangements of both nucleotides were observed in several crystal structures of complexes of the large ribosomal subunit with substrate analogs [10]. In particular, U2585 flexibility was suggested to provide means for the 3′ end tRNA A- to P-site rotatory motion [11], for D-amino acid rejection [12], and for facilitating the synergistic action of the streptogramins [13]. Hence, it appears that the binding of pleuromutilins utilizes the intrinsic flexibility of the ribosome, and thereby undergoes induced fit for binding of a large

C. Davidovich, *Targeting Functional Centers of the Ribosome*, Springer Theses, DOI: 10.1007/978-3-642-16931-1_4, © Springer-Verlag Berlin Heidelberg 2011

ensemble of compounds. Such induced-fit mechanism is consistent with kinetic data showing a prolonged retapamulin off-rate from *S. aureus* ribosomes, as well as with data suggesting that retapamulin inhibits P-site substrate binding independently of substrate concentration [14].

4.1.2 C14 Extension is Located in the PTC Void

In all cases, the C14 extension is located in the PTC and is held mainly by a two- to three–hydrogen-bond network between G2061 and the essential C21 keto of the bound compound [15]. In addition, in all cases U2585 is shifted away from the bound compound, to allow pleuromutilins binding. However, although in its altered location U2585 is in close proximity to C14 extension, it does not interact with it (Fig. 3.1d–f), regardless of the chemical nature of the C14 extension. The most prominent alteration of U2585 has been observed in the presence of the rigid C14 extension of SB-571519, where U2585 is shifted > 3 Å away from its native conformation (Fig. 3.1g).

In all three cases, the C14 extensions are located at the PTC, between the acetylated and the peptidylated tRNA CCA ends (Fig. 3.2b) [16]. All amino acids, of varying chemical characteristics, can be equally accommodated in this space. As this region contain only a few candidates for interactions, by its nature it should tolerate binding of various chemical moieties [17]. This explains why all C14 extensions of the bound compounds hardly interact with PTC components. Instead, longer extensions may interact with distal rRNA nucleotides, as observed crystallographically for tiamulin [5] or suggested, based on biochemical evidences, for valnemulin [16]. The small number of interactions of the C14 extensions rationalizes the similar activity of the different C14 pleuromutilin derivatives [18] and permits more flexibility in the design of this moiety.

Comparing the present findings with previous results [14, 15, 19] indicated that conditional to pleuromutilins activity is the H-bond network of C14 with G2061. Of importance is the C21 carbonyl group (Fig. 1.2) that exists also in the pleuromutilin mother compound and forms an H-bond with G2061 in all of the structurally studied pleuromutilins. Consistently, attempts to eliminate this group resulted in significantly lower or no activity [15, 19]. An example for a successful modification of C14 extension is the C14-acyl-carbamate pleuromutilin derivatives, here represented by SB-571519. The slightly decreased binding affinity of the acyl-carbamate derivative in comparison to the sulfanyl acetate derivatives (Table 3.1) may result from less optimal H-bond formed by the rigid acyl-carbamate group with G2061. This may lead to a slightly lower affinity of SB-571519 in respect to the studied sulfanyl acetate derivatives (Table 3.1). Importantly, it was shown that this slight decrease in activity can be tolerated clinically, in view of the low metabolism rates of the acyl-carbamates [19, 20] compared with the C14-sulfanyl-acetate derivatives, which are rapidly eliminated by cytochrome P450 [21].

4.1.3 Pleuromutilins Resistance

The slow stepwise manner of the appearance of pleuromutilin resistance is consistent with the finding that several pleuromutilin resistant strains include more than one mutation [4]. Furthermore, the majority of the nucleotides that are mutated for acquiring resistance do not directly interact with the bound compound, consistent with the essentiality of the interacting nucleotides for ribosomal function. It appears, therefore, that the strategy for acquiring resistance to antibiotics targeting the vicinity of the PTC is mutating nonconserved ribosomal components that do not belong to the essential portion of the PTC, but are involved in an interaction network with PTC nucleotides in a manner crucial for the organization of its functional conformation. Thus, it appears that these mutations reshape the conformation of this region, and/or alter its functional flexibility, in a fashion similar to allosteric rearrangements.

Likewise, the nonconserved region of L3 protein, residing in the vicinity of the PTC highly conserved nucleotides, enables such mechanism by mediating fine tuning of the PTC key nucleotides. Crystal structures of ribosome bound pleuromutilins confirm that the majority of pleuromutilin–rRNA interactions are between the tricyclic core and the binding pocket, and that most of these contacts are formed regardless of the chemical nature of their C14 extension (Fig. 3.1d–f). Furthermore, the addition of hydroxyl group at C2 in SB-571519 did not result in remarkably different binding orientation. Hence, it is not surprising that many pleuromutilin resistance mutations, identified in animal pathogens, involve nucleotides residing in the vicinity of the mutilin core, namely G2032, C2055, G2447, C2499, A2572, and U2504 [4]. A similar effect has been previously observed for several mutations in the nearby protein L3.

Although somewhat different in sequence, in all known structures of the large ribosomal subunit [8, 9, 22, 23] protein L3 penetrates deeply toward the PTC (Fig. 3.3a–c). Several mutations that cluster in a loop-like region of L3 at the proximity of the PTC induce resistance or reduce susceptibility to pleuromutilins [4] (Fig. 3.3b), presumably by altering the rRNA conformation or flexibility in the vicinity of the binding site [1, 5]. Similar resistance mechanisms were observed in yeasts [24, 25], where mutations in the corresponding L3 residues 255–257 [26] are responsible for anisomycin resistance.

In D50S/SB-571519 complex, Arg-144 (D. radiodurans numbering) of L3 interacts electrostatically with the phosphate of U2506 (Fig. 3.3a), a nucleotide playing a key role in defining pleuromutilins binding surface, thus showing that this L3 region is capable of interacting with PTC nucleotides involved in pleuromutilin binding. These particular electrostatic interactions may be specific to D. radiodurans, as the entire six amino acids loop-like region of L3 hosting Arg-144 displays rather high sequence diversity Fig. 3.3b). Nevertheless, the principle of reshaping a functional region composed of highly conserve nucleotides by using a nonconserved protein segment, is likely to be of a general character.

4.1.4 Pleuromutilins Selectivity Acquired by Remote Interactions

The key for clinical usage of antibiotics is their selectivity, namely the distinction between pathogens and their eukaryotic hosts. As the pleuromutilins bind to the highly or universally conserved PTC nucleotides [27], their selectivity is attributed to differences in the shape of the boundaries of the binding pocket induced by the non conserved nucleotides residing in its vicinity. Thus, pleuromutilins seem to discriminate between eubacteria and higher organisms in a fashion similar to that used for acquiring resistance, exploiting interactions between nucleotides that do not interact with the bound compounds.

An example for kingdom-related differences in the orientation of a major element in the binding pocket is nucleotide U2504. In the structure of the eubacterial large ribosomal subunit from *D. radiodurans* [8], in its complex with pleuromutilin [5], and in the eubacterial *E. coli* [22] and *T. thermophilus* [23, 28] 70S ribosomal complexes, this nucleotide points toward the pleuromutilin binding pocket (Fig. 3.3d), whereas in the large ribosomal subunit of the Archaeon *Haloarcula marismortui* [9] this nucleotide is tilted away from the binding site, presumably owing to its interactions with C2055. The difference in the identity of this nucleotide, namely cytosine in eubacteria but adenine in archaea (> 80%) and eukarya (> 96%) [27] seems to indirectly alter the binding site conformation, and consequently the affinity of pleuromutilins. In support of this hypothesis is the C2055A mutation that decreases the susceptibility of *Brachyspira* species to pleuromutilins [4]. An additional component that may promote drug discrimination is the significant sequence variability of protein L3 loop-like region that may trigger rRNA conformational differences among the pleuromutilin-binding pocket of ribosomes from different species.

4.2 Structural Basis for Cross-resistance Between Ribosomal PTC Antibiotics

4.2.1 Resistance to PTC Antibiotics is Frequently Acquired by Mutating Remote Nucleotides

Because the PTC is highly conserved, the mechanisms for acquiring resistance are based on altering the conformation and the flexibility of nucleotides residing remotely. This trend is represented by Fig. 3.6a, which shows that approximately half of the nucleotides mediating antibiotic resistance reside at distances > 6 Å from the affected bound drug. Because such distances are too long for direct interactions with the drugs, resistance mechanisms to PTC antibiotics appear to be assisted by additional nucleotides capable of forming remote interactions that, in turn, alter the conformation and the flexibility of the binding pocket surface.

Because altering the identity of PTC nucleotides in the immediate vicinity of the bound antibiotics is unfavorable, a common mechanism for acquiring resistance to PTC antibiotics is mediated by altering remote nucleotides (Fig. 3.6a). Indeed, proximity limitations would have severely limited the nucleotides pool, which in principle can be increased significantly (e.g., by approximately the power of three) with respect to the distance between the altered nucleotide and the antibiotic binding site. However, because of the requirement to trigger progressive conformational rearrangements, there is an upper limit for the nucleotides that can be included in the pool. Therefore, nucleotides residing at relatively large distances (> 10 Å) from the binding site are less likely to significantly alter its conformation. Hence, the advantage of increasing the pool of potential nucleotides by including nucleotides residing far from the binding site is compromised by their minute contribution to resistance, consistent with the existence of relatively fewer such cases. This effect can be somewhat compensated by altering more than a single nucleotide, as observed for pleuromutilins [4].

Resistance acquired by remote mutations has also been observed in other antibiotics families. For example, erythromycin resistance can result from mutations in r-proteins L4 and L22 [29, 30] which do not interact directly with the bound drug [31]. However, because resistance to macrolides can be acquired by alterations of nucleotides that interact with the drug, such as A2058, it seems that remotely acquired resistance to macrolides is an alternative mechanism, contrary to resistance to PTC antibiotics that is typically mediated by remote interactions.

4.2.2 U2504 at the Crossroad of Remote Mutations Networks that Hamper Binding of PTC Antibiotics

U2504, which plays pivotal roles in resistance to PTC antibiotics, is part of the binding pockets of phenicols [31], lincosamides [31, 32], pleuromutilins [5, 33] and oxazolidinones [34, 35] (Figs. 4.1 and 4.2a–r). Mutations of U2504 were shown to promote resistance to pleuromutilin in the veterinary pathogens *Brachyspira pilosicoli* and *Brachyspira hyodysenteriae* [4] and to linezolid in the archaeon *H. halobium* [36]. However, because it resides close to the PTC center, in the first layer of the PTC nucleotides that define the binding pocket surface, its alteration is expected to cause serious problems or be impossible. Consequently, altering neighboring nucleotides that can remotely affect 2504 may circumvent its essentiality. A mechanism by which U2504 is being perturbed by mutations of proximal nucleotides was suggested for tiamulin resistance [4]. This mechanism was later extended to be the general mechanism for resistance and selectivity of the pleuromutilin family, based on comparative crystallographic studies [33]. Likewise, linezolid resistance was shown to be acquired by a mutation in nucleotide G2576 (G2576U) [37, 38] that is located > 6 Å away from the bound drug [34], but can affect the conformation of nucleotide U2504.

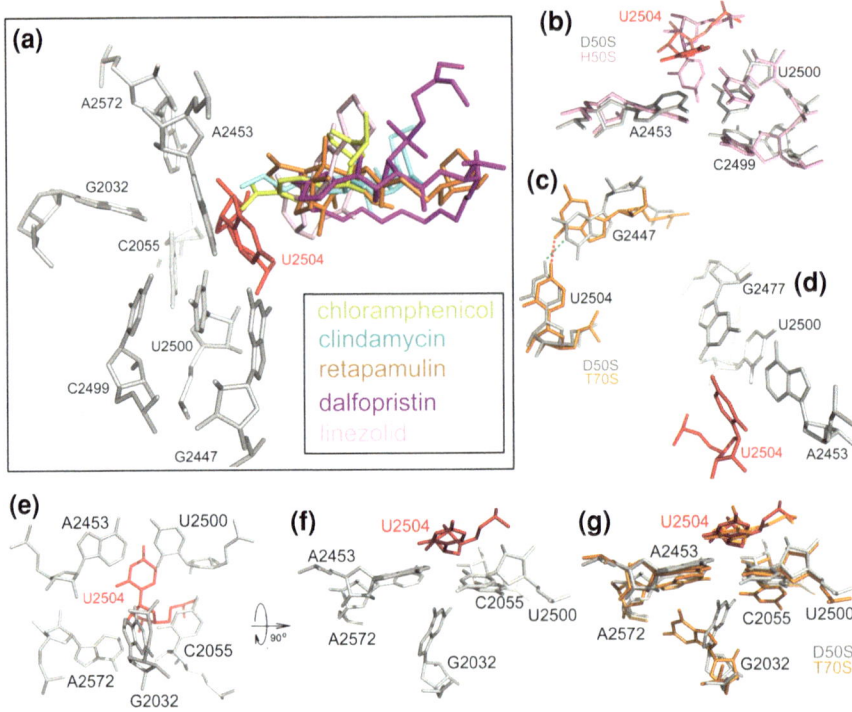

Fig. 4.1 Remote mutations that affect the conformation and/or the flexibility of U2504 by a network of interactions. **a** PTC antibiotics chloramphenicol (*yellow*), clindamycin (*cyan*), retapamulin (*orange*), dalfopristin (*magenta*) and linezolid (*pink*) bind in close proximity to U2504 (*red*). **b–g** Selected interactions within the networks around U2504. rRNA is *gray*, *pink*, or *orange* for D50S, H50S, or T70S, respectively. Wherever drawn, U2504 of D50S is *red*. Images of D50S, T70S, and H50S were generated from their coordinates (PDB ID codes 1NKW, 2J01, and 1S72, respectively)

Notably, in many cases, albeit not in all eubacteria, U2504 is a pseudouridine [39, 40]. Nevertheless, regardless of being uridine or pseudouridine, in all known structures of eubacterial ribosomes, U2504 makes similar interactions with its neighboring nucleotides (e.g., 2447) [8, 9, 22, 23, 28]. Interestingly, although this posttranscriptional modification is not necessary for smooth function of eubacterial ribosomes, it was linked to resistance to PTC antibiotics, because *E. coli* strains deficient of it are more susceptible to tiamulin, clindamycin, and linezolid [41]. The elevated susceptibility by the loss of posttranscriptional modification can be attributed, in part, to the centrality of this nucleotide in the binding pocket of PTC antibiotics. The same rationale, namely, the centrality of nucleotide U2504, explains why the key role played by this nucleotide in PTC resistance mechanism is independent of the presence or absence of the additional potential interactions that may accomplish by pseudouridine compared with unmodified nucleotide.

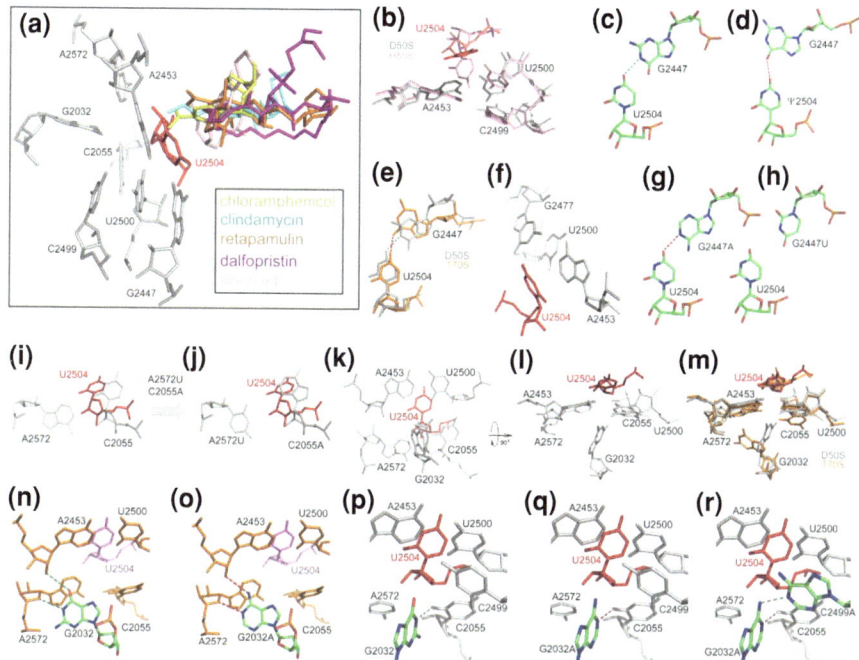

Fig. 4.2 Remote mutations that affect the conformation and/or the flexibility of U2504 by a network of interactions. **a** PTC antibiotics chloramphenicol (*yellow*), clindamycin (*cyan*), retapamulin (*orange*), dalfopristin (*magenta*), and linezolid (*pink*) bind in close proximity to U2504 (*red*). **b–r** The interactions network around U2504. rRNA is *gray*, *pink*, or *orange* for D50S, H50S, or T70S, respectively. Wherever indicated U2504 is *red* for D50S or *purple* for T70S. When applicable, different atoms are colored according to CPK color code to emphasize the chemical differences occurring by the mutagenesis. CPK colored nucleotides in *n* and *o* are of T70S, **d** shows E70S nucleotides after in silico modification from U to pseudo uracil at position 2504. All other images colored according to CPK are of D50S. Images of D50S, T70S, E70S, and H50S were generated from their coordinates (PDB ID codes 1NKW, 2J01, 2AWD, and 1S72, respectively)

4.2.3 Second Layer Nucleotides

The conformation and flexibility of nucleotide 2504 is dictated by five nucleotides residing within the second layer from the PTC wall, namely U2500, A2453, C2055, A2572 and G2447. The base pair U2500:A2453 forms a planar barrier that prevents U2504 from tilting away from the PTC [8, 23] (Figs. 4.1 and 4.2b). Thus, disruption of this pairing should release this planar barrier and enable U2504 to flip away from the PTC, in accord with the appearance of resistance to linezolid in clinical *S. aureus* isolates carrying a mutation U2500A [42] and in *H. halobium* strains carrying either of the induced mutations U2500C, A2453G, or A2453C [36], all of which are predicted to destabilize the U2500:A2453 base pair.

G2447 is an additional nucleotide residing in the second layer that forms close contacts (approximately 3 Å) with U2504 (Figs. 4.1c, d and 4.2c–h). In D50S its interactions with U2504 involve a hydrogen bond between O4 carbonyl of the uracil base of U2504 and N3 of G2447 (Figs. 4.1c and 4.2e). In the structures of *Thermus thermophilus* ribosomes (T70S) [23, 28] (Figs. 4.1c and 4.2e) and of *E. coli* (E70S) [22] O4 of U2504 faces the carbonyl O6 of G2447 base. In all structures G2447 stacks to U2500 (Figs. 4.1d and 4.2f) and thus restrains its conformation and indirectly facilitates further stabilization of U2504 by U2500. Hence, it is not surprising that although G2447 does not interact with any PTC antibiotics, its mutation G2447U (Fig. 4.2h) confers resistance to linezolid in *E. coli* [43] and *Mycobacterium smegmatis* [44], as well as to tiamulin in *E. coli* [37] and *B. hyodysenteriae* [4]. Furthermore, *E. coli* and *Bacillus stearothermophilus* ribosomes carrying the mutation G2447A (Fig. 4.2g) are not impaired by chloramphenicol in peptide bond formation in vitro [45], as positions 1 and 6 in this purine undergos a polar inversion in which both H bond donors become acceptors and vice versa. The conversion from purine to pyrimidine, followed the mutation G2447U, should result in exclusion of all interactions between these two nucleotides (Fig. 4.2h), as well as lose the stacking interaction with U2500.

Whereas U2500, A2453, and G2447 limit potential motions of U2504 base, both C2055 and A2572, located approximately 3 Å away from it, block its ribose sugar from shifting away from the PTC (Fig. 4.2i, j). The involvement of C2055 and A2572 in restraining U2504 conformation and flexibility is further supported by the tiamulin-resistant isolate of *B. hyodysenteriae* that carries both the mutations A2572U and C2055A [4] (Fig. 4.2j), although they are located 6–8 Å away from the bound drug. The purine to pyrimidine conversion after A2572U mutation releases the sterical hindrance of U2504 ribose sugar. In contrast, pyrimidine to purine conversion caused by the mutation C2055A can provide stacking interface that stabilizes U2504 in an altered conformation, with its base tilted away from the PTC, as observed for H50S [9], where nucleotide 2055 is an adenine. Thus, the different environments of U2504 in the eubacteria D50S and the archaea H50S, which is closer to eukaryotes in this aspect, explain the selectivity of pleuromutilins [33].

4.2.4 Third Layer Nucleotides

Alteration in the PTC surface can be performed also by third layer nucleotides. Among them nucleotide 2032, a highly conserved guanine in bacteria (> 94%), is involved in resistance to PTC antibiotics of four different families although it reside 6–8 Å from each of the bound drugs. Thus, the mutation G2032A (Figs. 4.1e–g and 4.2n–r) confers resistance to the antibiotics chloramphenicol, clindamycin [46], and linezolid [43] in *E. coli* but not in *T. thermophilus* [47]. The same mutation in *B. hyodysenteriae*, together with L3 r-protein mutation Asn148Ser, confers resistance to tiamulin [4]. In addition, elevated minimum inhibitory concentrations (MICs) of linezolid were observed for *E. coli* strains carrying the mutations G2032U and G2032C [43].

The base of G2032 in T70S is tilted by 90° with respect to the corresponding nucleotide in D50S (Figs. 4.1g and 4.2m). In T70S carbonyl O6 of G2032 forms a hydrogen bond with O2′ hydroxyl of nucleotide A2453 and its secondary amine N1 forms a hydrogen bond with O4′ of A2572 (Fig. 4.2n). G2032A mutation should lead to hydrogen bond acceptor imine at position N1 of adenine instead of the donor amine at position N1 of guanine. Additionally, the acceptor O6 carbonyl will be replaced by a primary amine that can act as a donor (Fig. 4.2o). These two polar inversions are likely to result in local repulsions that will force nucleotides A2453 and A2572 to adopt a slightly altered U2504 conformation. In D50S the secondary amine N1 of G2032 is located within a short distance of 2.5 Å from the pyrimidine ring carbonyl oxygen of the proximal C2055 (Fig. 4.2p). Hence, in the G2032A mutant the secondary amine hydrogen in position 1 of the guanine base is replaced by the imine lone pair of electrons located at the same position in adenine (Fig. 4.2q). This substitution can result in repulsion between the imine nitrogen of the mutated A2032 and the carbonyl oxygen of C2055 pyrimidine ring. These interactions are likely to indirectly perturb the conformation and flexibility of the proximal U2504.

C2499, a 100% conserved nucleotide in eubacteria, is also located in the third layer. Assuming that the conformation of 2032 in *B. hyodysenteriae* resembles its conformation in D50S, the C2499A mutation should stabilize the mutated nucleotide A2032 (originally guanine) by a favorable polar attraction between the primary amine at position 6 of A2032 and the imine nitrogen of the nearby mutated A2499 (originally cytosine) (Fig. 4.2r). The need for mutation in 2499 for the stabilization of 2032 in a conformation capable of pushing 2055 toward the PTC is further supported by the difference in the conformations of 2032 in T70S and D50S (Figs. 4.1g and 4.2m). It suggests that the unfavorable interactions with 2055, caused by the mutation G2032A without the compensating mutation C2499A, may stabilize 2032 in the T70S conformation. Therefore, the role of 2499 in resistance involves stabilizing the mutated 2032 in a conformation enabling repulsion of 2055, which ought to push 2504 toward the PTC and restrain its flexibility. Mutations in 2032 were observed in eubacteria, together with mutations of 2504 or 2499, as well as mutation in L3 Asn148Ser [4]. Whereas G2032A/U2504G double mutation should lead to resistance, mainly because of U2504G mutation, G2032A/C2499A should stimulate the mechanism proposed above [4]. The contribution of G2032A and C2499A double mutation for drug resistance in the later case is further supported by the reduced MIC of an identical bacterial strain that carries the same mutation in L3 protein but not the G2032A/C2499A double mutation [4]. The single example for mutation in 2032 causing linezolid resistance without the involvement of 2499 also fits with the general trait of resistance mediated by remote nucleotides.

The strategic position of U2504 at the A site and the cleft formed between it and the PTC rear wall serve as a hot spot for antibiotic binding (Figs. 4.1a and 4.2a), but is less suitable to mediate resistance owing to its high conservation.

However, U2504 possesses a significant level of flexibility, which seems to be used for acquiring pleuromutilin selectivity via remote interactions with less conserved components [33]. The similarities in the binding modes of all PTC antibiotics suggest that this mechanism can be extended to contribute to the selectivity of all five classes of clinically useful PTC antibiotics, in accord with the findings showing that the conformation of nucleotide 2504 affects the binding and resistance of PTC antibiotics even when the drugs are not in direct contact with it. Further support for this suggestion is the finding that 2504 acquires bacterial-like conformation in H50S when binding linezolid, despite the huge differences (three orders of magnitudes) in drug concentrations required for obtaining crystallographically suitable complexes of H50S [34] compared with D50S [35].

An additional nucleotide that can affect U2504 is its covalently attached neighbor A2503 (m^2A in *E. coli*) [48], which is prone to methylation in eubacterial strains carrying the resistance Cfr gene [2, 49, 50]. Supporting the suggestion that A2503 induces resistance indirectly is the finding that although chloramphenicol does not interact with A2503, alterations of A2503 lead to resistance probably through altered conformations of U2504 and G2061, in line with the finding that nucleotides that shape the antibiotics binding pockets determine the usefulness of bound antibiotics [51]; [52]; [71].

Although most of the resistance mechanisms mediated by indirect contacts seem to hinge on U2504, in a few cases PTC antibiotics are hindered remotely by other nucleotides. Among these are the mutation G2576U that was suggested to hamper linezolid binding indirectly by nucleotides U2505 and U2506 [34, 53] in addition to U2504, as well as the resistance to the streptogramin$_A$ virginiamycin M_1 in the archaeon *H. halobium* by mutation in nucleotide A2059 [54]. An open issue is the case of chloramphenicol, for which resistance by nucleotides 2057, 2058, and 2062, located at the exit tunnel, was reported [46, 55, 56] similar to chloramphenicol location in the ribosome of archaeon *H. marismortui* [32] that possesses sequence-resembling ribosomes from eukaryote, contrary to chloramphenicol binding site in the PTC of the pathogen model *D. radiodurans* [31].

4.2.5 Resistance to Various PTC Antibiotics Mediated by the Same Nucleotides

The involvement of the same nucleotides in resistance to several antibiotic families of different chemical natures (Table 2.2, Fig. 4.3, Figs. 3.4–3.6b) occurs presumably because of the overlapping binding sites of these drugs. As only a limited pool of nucleotides belonging to the PTC rear wall and the tunnel entrance is used for acquiring resistance, the probability of inducing resistance to more than a single antibiotic family by altering a given nucleotide is fairly high. This effect is further enhanced by the potential flexibility and the central location of U2504, which amplifies its possible involvement in resistance to various PTC antibiotics by indirect perturbation of its conformation and flexibility.

Resistance and/or reduced susceptibility to different antibiotics include (1) *E. coli* carrying the mutation C2032A that confers resistance to chloramphenicol and clindamycin [46]; (2) G2057A in other *E. coli* strains that causes resistance to chloramphenicol and erythromycin [55]; (3) A2058G and A2058U in *E. coli* acquiring reduced susceptibility to chloramphenicol and resistance to clindamycin and erythromycin [45]; (4) A2062C that appears together with the mutation Ser20Asn in L4 r-protein confirmed resistance to both streptogramins and macrolides in an isolate of *S. pneumoniae* [56].

Cross-resistance in the PTC was observed for linezolid and tiamulin in *S. aureus* following the mutation G2576U in all copies of 23S genes [37]. An example for potential clinical threat of cross-resistance in the PTC is the multidrug resistance phenotype mediated by the *cfr* rRNA methyltransferase. This gene encodes a methyltransferase that modifies the PTC nucleotide A2503, and is responsible for resistance to phenicols, lincosamides, oxazolidinones,

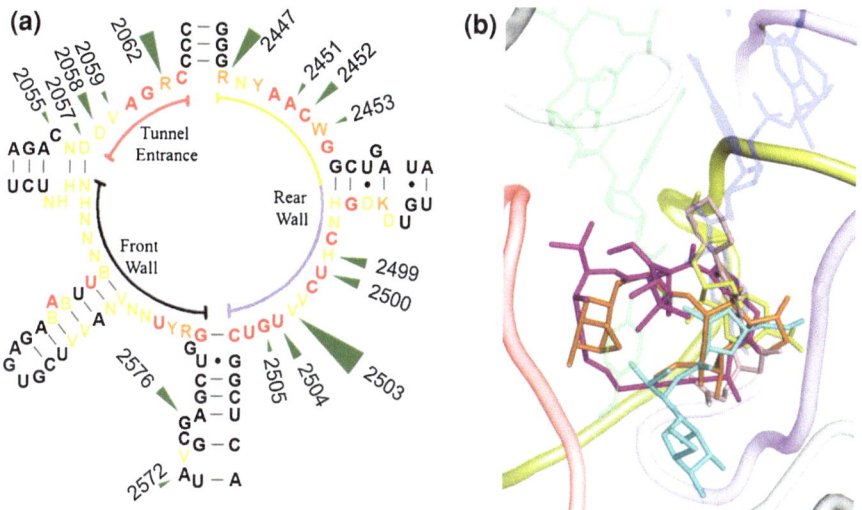

Fig. 4.3 Antibiotics binding pockets within the PTC. **a** 2D diagram of the 23S RNA at the vicinity of the PTC. *Green arrows* indicate nucleotides mediating resistance to PTC antibiotics. *Arrow* size is proportional to the number of different classes of antibiotics that are being affected. Relations between *E. coli* cell vitality and nucleotide alterations [70] are color coded. Absolutely essential nucleotide, which cannot be mutated are shown in *red*. Nucleotides that can be replaced by a single nucleotide are shown in *orange*. Nucleotides that can be replaced by 2 or 3 other nucleotides are shown in *yellow*. Nomenclature for Nucleotide Code Base is according to IUPAC [72]. Specific sections of the 2D diagram have colors identical to the colors of their corresponding regions in the 3D structure shown in **b**. **b** The Three-dimensional positions of PTC antibiotics showing their overlapping positions. The antibiotics chloramphenicol, clindamycin, retapamulin, dalfopristin, and linezolid are shown in *yellow*, *cyan*, *orange*, *magenta*, and *pink*, respectively. A site tRNA 3′ end and the derived P site tRNA are shown in transparent *blue* and *green*, respectively. The 23S rRNA is shown in *red*, *yellow*, and *blue*, as their corresponding sections in the 2D representation (**a**) 23S rRNA segments not shown in **a** are colored gray. The black arc in **a** is not shown in **b** because the latter is shown from the front wall direction

pleuromutilins, and streptogramin A antibiotics, namely the $PhLOPS_A$ group [2, 49]. Remarkably, a clinical isolate of *S. aureus* was shown to carry this gene together with the *ermB* gene on the same chromosome [50] and in the same operon [57]. The expression of this operon, designated *mlr* for modification of large ribosomal subunit, resulted in a strain resistant to all clinically relevant antibiotics that target the large ribosomal subunit.

To conclude, cross-resistance is of great importance in light of the increasing use of diverse antibiotics families hitting the same, or a similar, target. Structural studies on resistance mechanisms to various clinically useful families of PTC antibiotics revealed that all share overlapping positions in the vicinity of the PTC rear wall and nucleotide U2504. Common traits of resistance to several antibiotic families showed that almost all of the 23S nucleotides mediating resistance cluster within a defined small region of the PTC confined by its rear wall. Furthermore, approximately half of the nucleotides implicated in PTC antibiotic resistance are not directly interacting with the bound drugs, but reshape the binding pocket indirectly via networks of remote interactions, most of which through the flexible nucleotide U2504.

The significant number of nucleotides associated with multiple resistance phenotypes for PTC antibiotics indicates a linkage between structure, function, and resistance to several antibiotics, some with immense clinical value. Hence, the common traits of resistance revealed by this study may be useful in the design of preferred chemical moieties. Thus, it is likely that rather than being confined to the PTC rear wall, designing new PTC antibiotics with capabilities to dock to additional PTC components provides a feasible strategy for decreasing the probability for cross-resistance between PTC antibiotics.

4.3 Oligonucleotides as Ribosomal Inhibitors and as Tools for Structural and Functional Studies

The results suggested that the preferred target sites for short and long ODNs are characterized by different properties. It appears that short ODN displays lower IC_{50} when targeting rRNA sequences with relatively large accessible surface ($r = -0.65145$, $p = 0.057327$) and long 'base to base distance' ($r = -0.72791$, $p = 0.026194$), namely a rough measure for stacking, indicating the distance between N1 in pyrimidines and N9 in purines of two adjacent bases. These two parameters characterize accessible and poorly stacked rRNA. The results indicate that short oligonucleotides are preferred when target less conserved sequences, as was shown quantitatively by relatively low R_{seq} [58] ($r = 0.5244$, $p = 0.14724$). Such sites can be found on the periphery of the ribosome. These results indicating that short ODNs bind preferably to relatively exposed sites where they do not have to compensate for the high energetic penalty of perturbing the 3D structure in order to approach the target site. In case of short ODNs, this penalty may not be rewarded back upon hybridization. This can be further supported by the fact that most of the active short ODNs were designed to bind target sites that are clustering in the L7/12 stalk (Fig. 4.4). This ribosomal moiety is highly flexible and rather accessible in comparison with the majority of the well packed internal 23S rRNA.

IC_{50} in presence of the group of longer ODNs has poor correlation with r < 0.5 and p > 0.1 for all parameters, except for base to base distance (r = 0.53119, p = 0.075546). Thus, as of now solid conclusions regarding this group of ODNs, are not readily derived. However, the existence of statistically significant correlation between some of the tested parameters and the activity of long ODNs cannot be excluded. Hence, in order to derive such correlation coefficients, more observations are required (i.e. more ODNs to be tested).

A statistically valid (see Table 4.1) multiple linear model making chemical and biological sense could not be constructed when all ODNs were considered. The only valid models could be derived only if ODNs with length ≤ 15 were considered. Any attempt to include ODNs with higher length resulted with non valid models. This is in agreement with the diversity between the short and the long ODNs and with the relatively weak correlation between the long ODNs and the various parameters.

A total of 8,191 possible models were tested, considering all of the possible combinations of the best correlating parameters (n = 13). The selected model contains four parameters (see Table 4.1 for full description and Fig. 4.5 for predicted vs. observed IC_{50}). According to this model, an ODN is predicted to have a low IC_{50} if its target site is characterized by a relatively long base to base distances, high sequence conservation, low duplex formation energy and high target structure energy, namely the energy required to unwind the secondary structure of the rRNA at the target site. The model accords with chemical and biological sense. In addition it is mostly compatible with the trends observed when correlations were calculated for each parameter individually. The only parameter that appeared with different trend in the model is sequence conservation (mean R_{seq} [58]). When tested individually, this parameter was positively correlated with IC_{50} but when placed within the global model it displays a negative regression coefficient. This is presumably because the parameter serving as indicator for ribosomal periphery was first tested individually but pointed for highly functional centers when presented in the model.

4.4 Minimal Ribosomal Components with PTC Structure and Function

It is assumed that in the early terrestrial environment, the genetic information that was embedded in RNA sequences could lead to self replication and to phenotypes with catalytic properties [59–66]. In the case of the proto-ribosome, it is likely that the more efficient and more stable RNA dimers with pocket like structure that

Fig. 4.4 Location of the short target sites on the ribosome and activity of their complimentary ODNs. Target sites are presented as space filling models and are colored according to the IC_{50} of their corresponding ODNs (*green*: $IC_{50} < 40 \ \mu M$, *yellow*: $40 \ \mu M \le IC_{50} \le 80 \ \mu M$, *red*: $IC_{50} > 80 \ \mu M$, hybrid colors are representing overlapping target sites). **a** Front ("crown") view, **b** side view and **c** top view

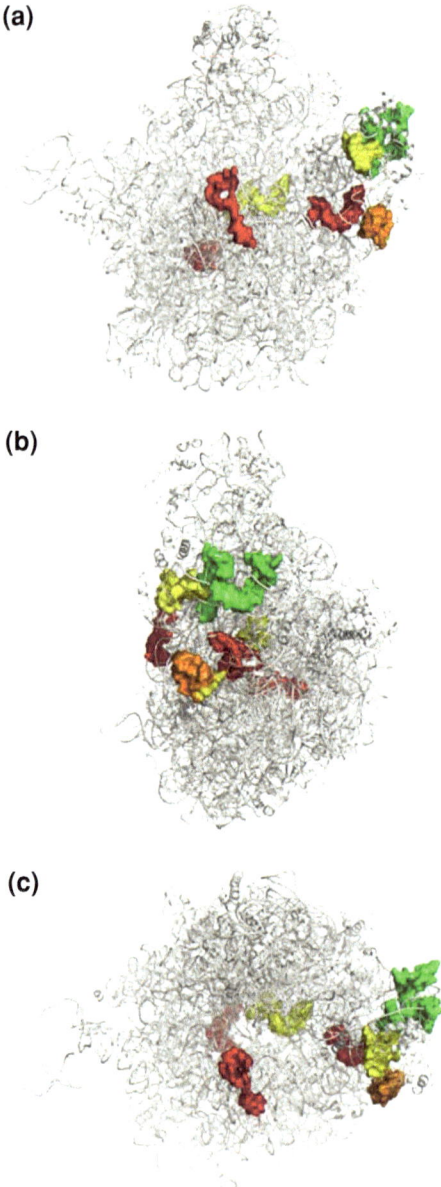

(a)

(b)

(c)

functioned as proto-ribosomes by positioning the substrates in a spatial arrangement facilitating peptide bond formation, had a better chance to be replicated. Thus, the surviving ancient pockets could become the templates for the ancient ribosomes. In a later stage these molecular entities underwent optimization from non-genetic peptide bond formation towards performing genetically driven translation.

Table 4.1 Multiple linear regression model for the prediction of IC_{50} values of short ODNs ($10 \leq$ length ≤ 15)

Parameter	Regression coefficient	Units	p Value
Average base-to-base distance (see text)	-61.2817 ± 13.4477	Å/base	0.001372
Sequence conservation (mean Rseq)	-154.5816 ± 43.5188	$base^{-1}$	0.006196
Duplex formation energy	6.9786 ± 2.3291	(kcal/mol)	0.015047
Target structure energy	-0.035185 ± 0.01048	(kcal/mol)	0.008425
Intercept estimate	899.7343	μM	

The model is based on all ODNs with length ≤ 15 (n = 14). Presented are regression coefficients and their standard error, as well as p values for testing whether a given regression coefficient is equal to zero. Overall p value for the model is 0.011445

The transition from a molecule forming peptide bonds to an elaborate apparatus capable of decoding genetic information seems to be coupled with the evolving genetic code and the proto-ribosome substrates. A higher level of efficiency could have been achieved by the creation of a supporting environment alongside the elongation of the substrates form the minimal mono- or oligonucleotide/amino acid conjugates, towards the modern tRNA. The conversion into longer compounds with a contour that can complement the inner surface of the reaction pocket occurred concurrently with proto-ribosome size increase, by the mutational optimization, aimed at accurate substrate positioning, which in the modern ribosome is governed by remote interactions between the RNA and the cavity leading to the PTC [11, 17]. This scenario accords with the fact that in the contemporary ribosome the symmetry is related to the backbone fold and not to the nucleotide sequence, thus emphasizing the superiority of functional requirement over sequence conservation.

A substantial increase in the catalytic rate could be obtained by the inclusion of peripheral elements, as observed for other ribozymes [67]. These additionally recruited structural elements could have been RNA chains or elongated oligopeptides that interacted with the proto-ribosome and its surroundings in a manner resembling the protein-RNA interactions in the modern ribosome. These could contribute to the suitability of the apparatus to act as an efficient machine and to support the emergence of genetic code translation. Mutational optimization that facilitated distinction between the two sides of the active site allowed differentiation between the two substrates, namely the acylated and the peptidyl tRNA. Besides functional optimization the nucleotide identities and conformations evolved for enhancing the stability of the symmetrical region. Consequently, the orientation of some RNA bases of the contemporary symmetrical region violate the internal symmetry (Fig. 4.6). Thus, with selection pressure for increased stability and efficiency, the proto-ribosome evolved into an entity that can provide all the activities required for nascent protein elongation in the modern ribosome. As such, the modern ribosome could have evolved gradually around the symmetrical

Fig. 4.5 Predicted vs. observed IC$_{50}$ values. Predicted values were calculated according to the multiple linear regression model suggested in Table 4.1. Observed values are as presented in Table 3.2

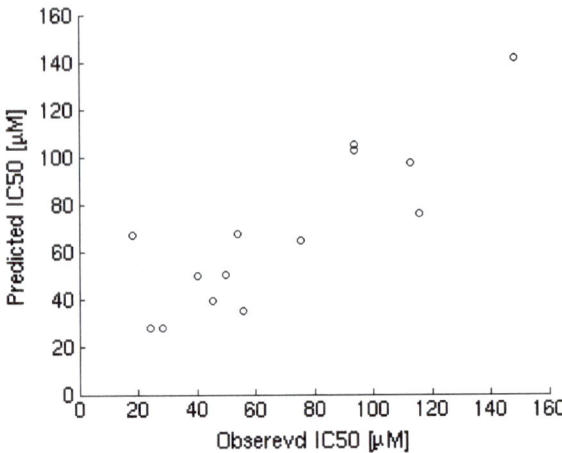

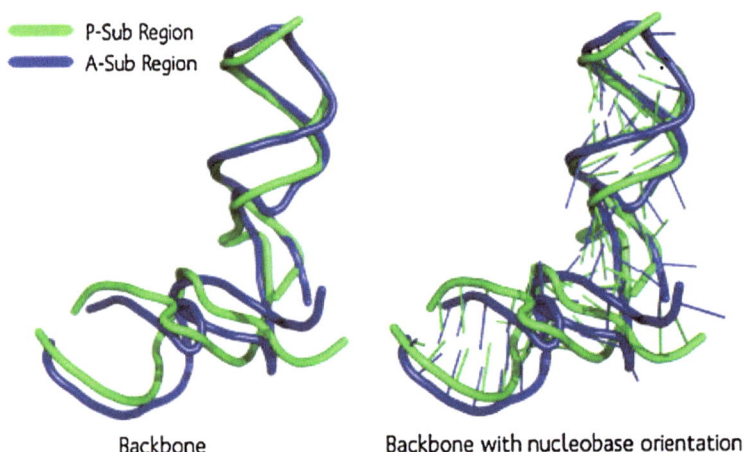

Backbone   Backbone with nucleobase orientation

Fig. 4.6 Superposition of the rRNA in the A- and the P-subregions (in *blue* and *green*, respectively). The striking overlap of the backbones of the two subregions, indicating the high level of symmetry, is shown on the left. Deviations from the symmetrical arrangements, indicated by the differences in the orientations of the non base-paired bases of the two subregions, are suggestive of the optimization that each half of the PTC underwent in order to fit the specific task of each site

region until it acquired its final shape and could perform programmed translation, either hierarchically [68] or by another mechanism (e.g. [69]).

The emergence of Life required an apparatus for synthesizing polypeptides capable of performing catalytic or other life supporting tasks, i.e. the ribosome. The proto-ribosome, which served as the precursor for the modern translation machinery by its capacity to autonomously catalyze peptide bonds forming non-coded amino acid oligo- or polymers, is suggested to have appeared by spontaneous dimeric assembly of two self-folding RNA chains. These pocket-like dimers offered a catalytic site for favorable positioning of the substrates involved in peptide bond formation and simple elongation. Assuming the existence of a proto ribosome my studies suggest that it is likely that it is still embedded in the core of the modern ribosome, and that the tendency for dimerization of the proto-ribosome, a prerequisite for obtaining the catalytic center, is intrinsically linked to its sequence thus indicating functional selection at the molecular level in the prebiotic era.

References

1. Bosling J, Poulsen SM, Vester B, Long KS (2003) Resistance to the peptidyl transferase inhibitor tiamulin caused by mutation of ribosomal protein l3. Antimicrob Agents Chemother 47(9):2892–2896
2. Long KS, Poehlsgaard J, Kehrenberg C, Schwarz S, Vester B (2006) The Cfr rRNA methyltransferase confers resistance to phenicols, lincosamides, oxazolidinones, pleuromutilins, and streptogramin A antibiotics. Antimicrob Agents Chemother 50(7):2500–2505
3. Poulsen SM, Karlsson M, Johansson LB, Vester B (2001) The pleuromutilin drugs tiamulin and valnemulin bind to the RNA at the peptidyl transferase centre on the ribosome. Mol Microbiol 41(5):1091–1099
4. Pringle M, Poehlsgaard J, Vester B, Long KS (2004) Mutations in ribosomal protein L3 and 23S ribosomal RNA at the peptidyl transferase centre are associated with reduced susceptibility to tiamulin in *Brachyspira spp.* isolates. Mol Microbiol 54(5):1295–1306
5. Schluenzen F, Pyetan E, Fucini P, Yonath A, Harms J (2004) Inhibition of peptide bond formation by pleuromutilins: the structure of the 50S ribosomal subunit from *Deinococcus radiodurans* in complex with tiamulin. Mol Microbiol 54(5):1287–1294
6. Youngman EM, Brunelle JL, Kochaniak AB, Green R (2004) The active site of the ribosome is composed of two layers of conserved nucleotides with distinct roles in peptide bond formation and peptide release. Cell 117(5):589–599
7. Hirabayashi N, Sato NS, Suzuki T (2006) Conserved loop sequence of helix 69 in *Escherichia coli* 23 S rRNA is involved in A-site tRNA binding and translational fidelity. J Biol Chem 281(25):17203–17211
8. Harms J, Schluenzen F, Zarivach R, Bashan A, Gat S, Agmon I, Bartels H, Franceschi F, Yonath A (2001) High resolution structure of the large ribosomal subunit from a mesophilic eubacterium. Cell 107(5):679–688
9. Ban N, Nissen P, Hansen J, Moore PB, Steitz TA (2000) The complete atomic structure of the large ribosomal subunit at 2.4 Å resolution. Science 289(5481):905–920
10. Schmeing TM, Huang KS, Strobel SA, Steitz TA (2005) An induced-fit mechanism to promote peptide bond formation and exclude hydrolysis of peptidyl-tRNA. Nature 438(7067):520–524
11. Bashan A, Agmon I, Zarivach R, Schluenzen F, Harms J, Berisio R, Bartels H, Franceschi F, Auerbach T, Hansen HAS, Kossoy E, Kessler M, Yonath A (2003) Structural basis of the ribosomal machinery for peptide bond formation, translocation, and nascent chain progression. Mol Cell 11:91–102
12. Agmon I, Amit M, Auerbach T, Bashan A, Baram D, Bartels H, Berisio R, Greenberg I, Harms J, Hansen HA, Kessler M, Pyetan E, Schluenzen F, Sittner A, Yonath A, Zarivach R (2004) Ribosomal crystallography: a flexible nucleotide anchoring tRNA translocation,

facilitates peptide-bond formation, chirality discrimination and antibiotics synergism. FEBS Lett 567(1):20–26

13. Harms J, Schluenzen F, Fucini P, Bartels H, Yonath A (2004) Alterations at the peptidyl transferase centre of the ribosome induced by the synergistic action of the streptogramins dalfopristin and quinupristin. BMC Biol 2:4

14. Yan K, Madden L, Choudhry AE, Voigt CS, Copeland RA, Gontarek RR (2006) Biochemical characterization of the interactions of the novel pleuromutilin derivative retapamulin with bacterial ribosomes. Antimicrob Agents Chemother 50(11):3875–3881

15. Egger H, Reinshagen H (1976) New pleuromutilin derivatives with enhanced antimicrobial activity. II. Structure-activity correlations. J Antibiot (Tokyo) 29(9):923–927

16. Nissen P, Kjeldgaard M, Nyborg J (2000) Macromolecular mimicry. EMBO J 19(4):489–495

17. Yonath A (2003) Ribosomal tolerance and peptide bond formation. Biol Chem 384(10): 1411–1419

18. Berry V, Dabbs S, Frydrych CH, Hunt E, Woodnutt G, Sanderson FD (1999) Pleuromutilin derivatives as antimicrobials. Patent number WO9921855

19. Brooks G, Burgess W, Colthurst D, Hinks JD, Hunt E, Pearson MJ, Shea B, Takle AK, Wilson JM, Woodnutt G (2001) Pleuromutilins. Part 1. The identification of novel mutilin 14-carbamates. Bioorg Med Chem 9(5):1221–1231

20. Hunt E (2000) Pleuromutilin antibiotics. Drugs Future 25(11):1163–1168

21. Schuster I, Fleschurz C, Hildebrandt J, Turnowsky F, Zsutty H, Kretschmer G, Spitzy KH, Karrer K (1983) Binding and degradation of TDM 85-530 by a microsomal Cyt P-450 form from man, rat and mouse in vitro. Proceedings of 13th International Congress of Chemotherapy, vol 5, pp 108/142–108/146

22. Schuwirth BS, Borovinskaya MA, Hau CW, Zhang W, Vila-Sanjurjo A, Holton JM, Cate JHD (2005) Structures of the bacterial ribosome at 3.5 Å resolution. Science 310(5749):827–834

23. Selmer M, Dunham CM, FVt Murphy, Weixlbaumer A, Petry S, Kelley AC, Weir JR, Ramakrishnan V (2006) Structure of the 70S ribosome complexed with mRNA and tRNA. Science 313(5795):1935–1942

24. Fried HM, Warner JR (1981) Cloning of yeast gene for trichodermin resistance and ribosomal protein L3. Proc Natl Acad Sci USA 78(1):238–242

25. Jimenez A, Sanchez L, Vazquez D (1975) Simultaneous ribosomal resistance to trichodermin and anisomycin in *Saccharomyces cerevisiae* mutants. Biochim Biophys Acta 383(4):427–434

26. Meskauskas A, Petrov AN, Dinman JD (2005) Identification of functionally important amino acids of ribosomal protein L3 by saturation mutagenesis. Mol Cell Biol 25(24):10863–10874

27. Cannone JJ, Subramanian S, Schnare MN, Collett JR, D'Souza LM, Du Y, Feng B, Lin N, Madabusi LV, Müller KM, Pande N, Shang Z, Yu N, Gutell RR (2002) The comparative RNA Web (CRW) site: an online database of comparative sequence and structure information for ribosomal, intron, and other RNAs. BMC Bioinformatics 3(2):1–31

28. Korostelev A, Trakhanov S, Laurberg M, Noller HF (2006) Crystal structure of a 70S ribosome-tRNA complex reveals functional interactions and rearrangements. Cell 126:1065–1077

29. Gregory ST, Dahlberg AE (1999) Erythromycin resistance mutations in ribosomal proteins L22 and L4 perturb the higher order structure of 23 S ribosomal RNA. J Mol Biol 289(4):827–834

30. Zaman S, Fitzpatrick M, Lindahl L, Zengel J (2007) Novel mutations in ribosomal proteins L4 and L22 that confer erythromycin resistance in *Escherichia coli*. Mol Microbiol 66(4):1039–1050

31. Schluenzen F, Zarivach R, Harms J, Bashan A, Tocilj A, Albrecht R, Yonath A, Franceschi F (2001) Structural basis for the interaction of antibiotics with the peptidyl transferase centre in eubacteria. Nature 413(6858):814–821

32. Hansen JL, Moore PB, Steitz TA (2003) Structures of five antibiotics bound at the peptidyl transferase center of the large ribosomal subunit. J Mol Biol 330(5):1061–1075

33. Davidovich C, Bashan A, Auerbach-Nevo T, Yaggie RD, Gontarek RR, Yonath A (2007) Induced-fit tightens pleuromutilins binding to ribosomes and remote interactions enable their selectivity. Proc Natl Acad Sci USA 104(11):4291–4296

34. Ippolito JA, Kanyo ZF, Wang D, Franceschi FJ, Moore PB, Steitz TA, Duffy EM (2008) Crystal structure of the oxazolidinone antibiotic linezolid bound to the 50S ribosomal subunit. J Med Chem 51(12):3353–3356

35. Wilson DN, Schluenzen F, Harms JM, Starosta AL, Connell SR, Fucini P (2008) The oxazolidinone antibiotics perturb the ribosomal peptidyl-transferase center and effect tRNA positioning. Proc Natl Acad Sci USA 105(26):13339–13344

36. Kloss P, Xiong L, Shinabarger DL, Mankin AS (1999) Resistance mutations in 23 S rRNA identify the site of action of the protein synthesis inhibitor linezolid in the ribosomal peptidyl transferase center. J Mol Biol 294(1):93–101

37. Miller K, Dunsmore CJ, Fishwick CW, Chopra I (2008) Linezolid and tiamulin cross-resistance in *Staphylococcus aureus* mediated by point mutations in the peptidyl-transferase center. Antimicrob Agents Chemother 52(5):1737–1742

38. Prystowsky J, Siddiqui F, Chosay J, Shinabarger DL, Millichap J, Peterson LR, Noskin GA (2001) Resistance to linezolid: characterization of mutations in rRNA and comparison of their occurrences in vancomycin-resistant enterococci. Antimicrob Agents Chemother 45(7):2154–2156

39. Del Campo M, Recinos C, Yanez G, Pomerantz SC, Guymon R, Crain PF, McCloskey JA, Ofengand J (2005) Number, position, and significance of the pseudouridines in the large subunit ribosomal RNA of *Haloarcula marismortui* and *Deinococcus radiodurans*. RNA 11(2):210–219

40. Ofengand J, Bakin A (1997) Mapping to nucleotide resolution of pseudouridine residues in large subunit ribosomal RNAs from representative eukaryotes, prokaryotes, archaebacteria, mitochondria and chloroplasts. J Mol Biol 266(2):246–268

41. Toh SM, Mankin AS (2008) An indigenous posttranscriptional modification in the ribosomal peptidyl transferase center confers resistance to an array of protein synthesis inhibitors. J Mol Biol 380(4):593–597

42. Meka VG, Pillai SK, Sakoulas G, Wennersten C, Venkataraman L, DeGirolami PC, Eliopoulos GM, Moellering RC Jr, Gold HS (2004) Linezolid resistance in sequential *Staphylococcus aureus* isolates associated with a T2500A mutation in the 23S rRNA gene and loss of a single copy of rRNA. J Infect Dis 190(2):311–317

43. Xiong L, Kloss P, Douthwaite S, Andersen NM, Swaney S, Shinabarger DL, Mankin AS (2000) Oxazolidinone resistance mutations in 23S rRNA of *Escherichia coli* reveal the central region of domain V as the primary site of drug action. J Bacteriol 182(19):5325–5331

44. Sander P, Belova L, Kidan YG, Pfister P, Mankin AS, Bottger EC (2002) Ribosomal and non-ribosomal resistance to oxazolidinones: species-specific idiosyncrasy of ribosomal alterations. Mol Microbiol 46(5):1295–1304

45. Thompson J, Kim DF, O'Connor M, Lieberman KR, Bayfield MA, Gregory ST, Green R, Noller HF, Dahlberg AE (2001) Analysis of mutations at residues A2451 and G2447 of 23S rRNA in the peptidyltransferase active site of the 50S ribosomal subunit. Proc Natl Acad Sci USA 98(16):9002–9007

46. Douthwaite S (1992) Interaction of the antibiotics clindamycin and lincomycin with *Escherichia coli* 23S ribosomal RNA. Nucleic Acids Res 20(18):4717–4720

47. Gregory ST, Carr JF, Rodriguez-Correa D, Dahlberg AE (2005) Mutational analysis of 16S and 23S rRNA genes of *Thermus thermophilus*. J Bacteriol 187(14):4804–4812

48. Kowalak JA, Bruenger E, McCloskey JA (1995) Posttranscriptional modification of the central loop of domain V in *Escherichia coli* 23 S ribosomal RNA. J Biol Chem 270(30):17758–17764

49. Kehrenberg C, Schwarz S, Jacobsen L, Hansen LH, Vester B (2005) A new mechanism for chloramphenicol, florfenicol and clindamycin resistance: methylation of 23S ribosomal RNA at A2503. Mol Microbiol 57(4):1064–1073

50. Toh SM, Xiong L, Arias CA, Villegas MV, Lolans K, Quinn J, Mankin AS (2007) Acquisition of a natural resistance gene renders a clinical strain of methicillin-resistant *Staphylococcus aureus* resistant to the synthetic antibiotic linezolid. Mol Microbiol 64(6):1506–1514

51. Bommakanti AS, Lindahl L, Zengel JM (2008) Mutation from guanine to adenine in 25S rRNA at the position equivalent to *E. coli* A2058 does not confer erythromycin sensitivity in *Sacchromyces cerevisae*. RNA 14(3):460–464

52. Yonath A (2005) Antibiotics targeting ribosomes: resistance, selectivity, synergism, and cellular regulation. Annu Rev Biochem 74:649–679

53. Miller VM, Paulson HL, Gonzalez-Alegre P (2005) RNA interference in neuroscience: progress and challenges. Cell Mol Neurobiol 25(8):1195–1207

54. Porse BT, Garrett RA (1999) Sites of interaction of streptogramin A and B antibiotics in the peptidyl transferase loop of 23 S rRNA and the synergism of their inhibitory mechanisms. J Mol Biol 286(2):375–387

55. Ettayebi M, Prasad SM, Morgan EA (1985) Chloramphenicol-erythromycin resistance mutations in a 23S rRNA gene of *Escherichia coli*. J Bacteriol 162(2):551–557

56. Mankin AS, Garrett RA (1991) Chloramphenicol resistance mutations in the single 23S rRNA gene of the archaeon *Halobacterium halobium*. J Bacteriol 173(11):3559–3563

57. Smith LK, Mankin AS (2008) Transcriptional and translational control of the mlr operon which confers resistance to seven classes of protein synthesis inhibitors. Antimicrob Agents Chemother 52(5):1703–1712

58. Schneider TD, Stormo GD, Gold L, Ehrenfeucht A (1986) Information content of binding sites on nucleotide sequences. J Mol Biol 188(3):415–431

59. Eigen M, Lindemann BF, Tietze M, Winkler-Oswatitsch R, Dress A, von Haeseler A (1989) How old is the genetic code? Statistical geometry of tRNA provides an answer. Science 244(4905):673–679

60. Ferris JP (2002) Montmorillonite catalysis of 30–50 mer oligonucleotides: laboratory demonstration of potential steps in the origin of the RNA world. Orig Life Evol Biosph 32(4):311–332

61. Lincoln TA, Joyce GF (2009) Self-sustained replication of an RNA enzyme. Science 323(5918):1229–1232

62. Pino S, Ciciriello F, Costanzo G, Di Mauro E (2008) Nonenzymatic RNA ligation in water. J Biol Chem 283(52):36494–36503

63. Smith JM, Szathmáry E (1995) The major transitions in evolution. Oxford University Press, New York

64. Woese CR (1973) The rotating ribosome: a gross mechanical model for translation. J Theor Biol 38(1):203–204

65. Woese CR (2001) Translation: in retrospect and prospect. RNA 7(8):1055–1067

66. Yarus M (2002) Primordial genetics: phenotype of the ribocyte. Annu Rev Genet 36:125–151

67. Voytek SB, Joyce GF (2007) Emergence of a fast-reacting ribozyme that is capable of undergoing continuous evolution. Proc Natl Acad Sci USA 104(39):15288–15293

68. Bokov K, Steinberg SV (2009) A hierarchical model for evolution of 23S ribosomal RNA. Nature 457(7232):977–980

69. Fox GE, Naik AK (2004) The evolutionary history of the ribosome. In: Ribas de Poplana L (ed) The genetic code and the origin of life. Landes Bioscience, Austin, pp 92–105

70. Sato NS, Hirabayashi N, Agmon I, Yonath A, Suzuki T (2006) Comprehensive genetic selection revealed essential bases in the peptidyl-transferase center. Proc Natl Acad Sci USA 103(42):15386–15391

71. Yonath A, Bashan A (2004) Ribosomal crystallography: Initiation, peptide bond formation, and amino acid polymerization are hampered by antibiotics. Annu Rev Microbiol 58(380): 233–251

72. Cornish-Bowden A (1985) Nomenclature for incompletely specified bases in nucleic acid sequences: recommendations 1984. Nucleic Acids Res 13(9):3021–3030